Buckle Down™

to the Common Core State Standards

Mathematics

Grade 4

This book belongs to: ______________________________

Helping your schoolhouse meet the standards of the statehouse™

ISBN 978-0-7836-7986-0

1CCUS04MM01 10

Cover Image: Colorful marbles. © Corbis/Photolibrary

Triumph Learning® 136 Madison Avenue, 7th Floor, New York, NY 10016

Printed in the United States of America.

Frequently Asked Questions about the Common Core Standards

What are the Common Core State Standards?

The Common Core State Standards for mathematics and English language arts, grades K–12, are a set of shared goals and expectations for the knowledge and skills that will help students succeed. They allow students to understand what is expected of them and to become progressively more proficient in understanding and using mathematics and English language arts. Teachers will be better equipped to know exactly what they must do to help students learn and to establish individualized benchmarks for them.

Will the Common Core State Standards tell teachers how and what to teach?

No. Because the best understanding of what works in the classroom comes from teachers, these standards will establish *what* students need to learn, but they will not dictate *how* teachers should teach. Instead, schools and teachers will decide how best to help students reach the standards.

What will the Common Core State Standards mean for students?

The standards will provide a clear, consistent understanding of what is expected of student learning across the country. Common standards will not prevent different levels of achievement among students, but they will ensure more consistent exposure to materials and learning experiences through curriculum, instruction, teacher preparation, and other supports for student learning. These standards will help give students the knowledge and skills they need to succeed in college and careers.

Do the Common Core State Standards focus on skills and content knowledge?

Yes. The Common Core State Standards recognize that both content and skills are important. They require rigorous content and application of knowledge through higher-order thinking skills. The English language arts standards require certain critical content for all students, including classic myths and stories from around the world, America's founding documents, foundational American literature, and Shakespeare. The remaining crucial decisions about content are left to state and local determination. In addition to content coverage, the Common Core State Standards require that students systematically acquire knowledge of literature and other disciplines through reading, writing, speaking, and listening.

In mathematics, the Common Core State Standards lay a solid foundation in whole numbers, addition, subtraction, multiplication, division, fractions, and decimals. Together, these elements support a student's ability to learn and apply more demanding math concepts and procedures.

The Common Core State Standards require that students develop a depth of understanding and ability to apply English language arts and mathematics to novel situations, as college students and employees regularly do.

Will common assessments be developed?

It will be up to the states: some states plan to come together voluntarily to develop a common assessment system. A state-led consortium on assessment would be grounded in the following principles: allowing for comparison across students, schools, districts, states and nations; creating economies of scale; providing information and supporting more effective teaching and learning; and preparing students for college and careers.

TABLE OF CONTENTS

To the Teacher:

Standards Name numbers are listed for each lesson in the table of contents. The numbers in the shaded gray bar that runs across the tops of the pages in the workbook indicate the Standards Name for a given page (see example to the left).

Introduction

When you start thinking about all of the times and places you use your math skills, you might be surprised. How many minutes are left until school is over for the day? How much taller are you now than you were a year ago? How many points per game is the leading rusher for your favorite football team averaging? These are examples of how people use math every day.

This book will help you practice your everyday math skills. It will also help you with math that is a bit more unusual, like the kinds you often see in math class and on math tests. Since you already know quite a bit of math, practice will make you an even better math student.

Test-Taking Tips

Here are a few tips that will help you on test day.

TIP 1: Take it easy.

Stay relaxed and confident. Because you've practiced the problems in your workbook, you will be ready to do your best on almost any math test. Take a few slow, deep breaths before you begin the test.

TIP 2: Have the supplies you need.

For most math tests, you will need two sharp pencils and an eraser. Your teacher will tell you whether you need anything else.

TIP 3: Read the questions more than once.

Every question is different. Some questions are more difficult than others. If you need to, read a question more than once. This will help you make a plan for solving the question.

TIP 4: Learn to "plug in" answers to multiple-choice items.

When do you "plug in"? You should "plug in" whenever your answer is different from all of the answer choices or you can't come up with an answer. Plug each answer choice into the problem and find the one that makes sense. (You can also think of this as "working backward.")

TIP 5: Answer open-ended items completely.

When answering short-response and extended-response items, show all your work to receive as many points as possible. Write neatly enough so that your calculations will be easy to follow. Make sure your answer is clearly marked.

TIP 6: Use all the test time.

Work on the test until you are told to stop. If you finish early, go back through the test and double-check your answers. You just might increase your score on the test by finding and fixing any errors you might have made.

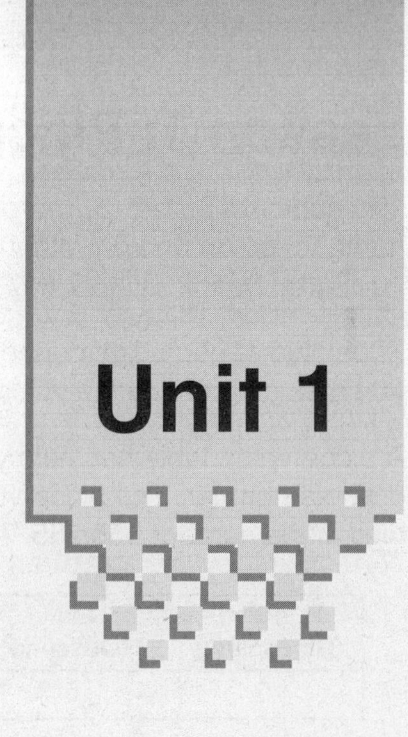

Number and Operations in Base Ten

You may not realize it, but you solve problems using number and operations in everyday thinking and activities many times a day. You use number and operations when you figure out how much time you have to get ready for school, about how much money you will need for a movie ticket and a box of popcorn, or whether it will be warm enough to go swimming. Number and operations are important in almost every part of your life.

In this unit, you will use place value and base-ten numerals to represent, compare, round, add, and subtract whole numbers. You will review estimation and learn some strategies for mentally calculating sums and differences.

In This Unit

Place Value and Whole Numbers

Using Place Value to Compare Whole Numbers

Rounding and Estimating with Whole Numbers

Adding Whole Numbers

Subtracting Whole Numbers

Lesson 1: Place Value and Whole Numbers

The numbers 0, 1, 2, 3,... are called **whole numbers**. The three dots show that the numbers go on and on without end. Our system of representing numbers uses 10 **digits**: 0, 1, 2, 3, 4, 5, 6, 7, 8, 9.

The value of each digit depends on its place in a numeral. The value of each place is 10 times as great as the value of the place next to it on the right.

A place-value table can help you recognize the value of each digit used in representing a whole number. The place-value table below shows the value represented by each digit in the number 555,555.

Hundred Thousands	Ten Thousands	Thousands	Hundreds	Tens	Ones
5	5	5	5	5	5
↑ The value of this 5 is 5 hundred thousands, or 500,000.	↑ The value of this 5 is 5 ten thousands, or 50,000.	↑ The value of this 5 is 5 thousands, or 5000.	↑ The value of this 5 is 5 hundreds, or 500.	↑ The value of this 5 is 5 tens, or 50.	↑ The value of this 5 is 5 ones, or 5.

You can represent a whole number in different ways.

You can use **expanded form** to show the value of each digit:

500,000 + 50,000 + 5000 + 500 + 50 + 5

You can use words to write or say the **number name**:

five hundred fifty-five thousand, five hundred fifty-five

You can use a **base-ten numeral**, or **standard form**, to write the number:

555,555

When numbers are represented by base-ten numerals, commas separate the digits of greater numbers so they can be read more easily. Starting with the digit at the far right and moving from right to left, a comma is placed to the left of every third digit in any number that is represented by 5 or more digits.

2 3,1 8 4 ← start

3 2 1

Example

Write the number represented in the place-value table below.
Use expanded form, the number name, and the base-ten numeral.

Hundred Thousands	Ten Thousands	Thousands	Hundreds	Tens	Ones
8	4	2	1	3	6

Expanded form: 800,000 + 40,000 + 2000 + 100 + 30 + 6
Number name: eight hundred forty-two thousand, one hundred thirty-six
Base-ten numeral: 842,136

Example

Write the base-ten numeral 308,041 in a place-value table.
Then write the expanded form and the number name.

Hundred Thousands	Ten Thousands	Thousands	Hundreds	Tens	Ones
3	0	8	0	4	1

Notice that there are no ten thousands or hundreds.
This is why a 0 (zero) is written in those places.
Zeros are not shown in expanded form or in the number name.
300,000 + 8000 + 40 + 1 = 308,041
3**0**8,**0**41 = three hundred eight thousand, forty-one

There are no ten thousands and no hundreds.

Practice

Use the top row of the place-value table to help you answer questions 1 through 3.

Hundred Thousands	Ten Thousands	Thousands	Hundreds	Tens	Ones

1. 100 tens is the same as how many thousands? ____________________

2. 10 thousands is the same as how many hundreds? ____________________

3. 1 hundred is the same as how many tens? ____________________

4. Write the digits for each of the following numbers in the correct places in the place-value table.

   48,104 289,726 42,947 900,015

Hundred Thousands	Ten Thousands	Thousands	Hundreds	Tens	Ones

5. What digit is in the ten-thousands place of 92,136? ________________

6. What digit is in the ones place of 4075? ________________

7. What digit is in the thousands place of 18,372? ________________

8. What digit is in the hundred-thousands place of 780,941? ________________

9. In which of the following base-ten numerals does the digit in the thousands place represent a value that is 10 times the value represented by the digit in the hundreds place?

   A. 2372

   B. 4889

   C. 5752

   D. 6671

10. In which of the following base-ten numerals does the digit in the tens place represent a value that is 10 times the value represented by the digit in the ones place?

    A. 43,755

    B. 44,821

    C. 46,630

    D. 47,172

11. The population of Elm City in a recent year was estimated to be 41,382.

    Write 41,382 in expanded form.

    ____

12. The distance from Ridgeville to Hamilton is 1075 miles. Write 1075 in word form.

    ____

13. Blue Ridge State Park covers approximately 2364 acres of land. Write 2364 in expanded form.

    ____

14. Eagle Hawk Mountain is 4801 feet above sea level. Write the number name for 4801.

    ____

15. Last year, 910,356 people visited the zoo. Write 910,356 in expanded form.

    ____

16. In the base-ten numeral 777, how is the value of the middle digit related to the values of the digits on the right and on the left?

    ____

    ____

    ____

Lesson 2: Using Place Value to Compare Whole Numbers

When you compare two numbers, you decide which number is **greater than** the other or which number is **less than** the other. The signs used to compare numbers are >, <, and =.

> means **is greater than**

< means **is less than**

= means **is equal to**

You can use a place-value table to help compare whole numbers.
Start with the digit at the far left. Compare the digits in each place from left to right and look for the first place in which the digits are different.

Example

Compare the numbers 275,428 and 275,961.

The following place-value table shows that the numbers have the same digits in the hundred-thousands, ten-thousands, and thousands places. The first place in which the digits are different is the hundreds place.

Hundred Thousands	Ten Thousands	Thousands	Hundreds	Tens	Ones
2	7	5	4	2	8
2	7	5	9	6	1

The digit 4 represents 400, the digit 9 represents 900, and 400 is less than 900. So 275,428 is less than 275,961.
Write: 275,428 < 275,961

It is also true that 900 is greater than 400, so 275,961 is greater than 275,428.
Write: 275,961 > 275,428

Example

Compare the numbers 65,847 and 65,837.
Compare the ten thousands: The ten thousands are the same.
Compare the thousands: The thousands are the same.
Compare the hundreds: The hundreds are the same.
Compare the tens: 4 represents 40, 3 represents 30, and 40 > 30.
So 65,847 > 65,837.

TIP: The wider side of the < or > symbol is always next to the greater number. The small, pointed side of the symbol always points to the smaller number.

Practice

Directions: For questions 1 through 10, use the table below to compare the numbers. Write the correct symbol (>, <, or =) on the blank.

Hundred Thousands	Ten Thousands	Thousands	Hundreds	Tens	Ones
	4	8	5	2	6
	4	8	7	0	1
	4	9	3	8	5
	4	8	4	9	3
	4	9	3	9	8
	4	9	0	7	2

1. 48,526 ________ 48,493
2. 48,701 ________ 48,526
3. 48,493 ________ 48,701
4. 48,493 ________ 49,385
5. 49,385 ________ 49,398
6. 49,072 ________ 49,398
7. 49,072 ________ 48,701
8. 48,526 ________ 49,072
9. 48,493 ________ 49,072
10. 49,385 ________ 49,385

Directions: For questions 11 through 14, use the table below to compare the numbers. Write the correct symbol (>, <, or =) on the blank.

Hundred Thousands	Ten Thousands	Thousands	Hundreds	Tens	Ones
1	1	0	4	5	6
	9	9	8	1	0
1	1	2	1	2	4
1	0	1	2	4	2
1	1	0	6	9	8

11. 99,810 ________ 110,456
12. 112,124 ________ 101,242
13. 110,698 ________ 112,124
14. 110,456 ________ 110,698

Directions: For questions 15 through 28, use >, <, or = to compare the numbers.

15. 742 __=__ 742
16. 810 __>__ 180
17. 2044 __<__ 4022
18. 10,539 __<__ 10,695
19. 28,513 __>__ 28,315
20. 31,450 __7__ 30,450
21. 38,844 __<__ 48,844
22. 71,204 __<__ 71,240
23. 67,246 __<__ 76,264
24. 40,892 __7__ 4621
25. 87,118 __<__ 87,119
26. 327,843 __<__ 372,843
27. 99,998 __<__ 100,001
28. 110,658 __<__ 120,658

Directions: Use the table below to answer question 29.

Pizzas Sold Last Year

Restaurant	Number of Pizzas
Pete's Pizzas	11,841
The Pizza Parlor	10,903
Pizza Heaven	9772
Betsy's Best Pizzas	10,655
Family Pizza	9323

29. Paula's Pizza Place sold 9780 pizzas last year. Explain how to find the restaurants that sold more pizzas than Paula's Pizza Place.

you look at the ten thoasands Place and which ever humber is bigger sold more pazza

Lesson 3: Rounding and Estimating with Whole Numbers

Estimation helps you find what number the answer to a problem should be close to. **Rounding** is the most common way to estimate. You can use place value to round a whole number to the nearest ten, hundred, thousand, ten thousand, or hundred thousand.

To round a whole number to any place:

- Circle the digit in the place to be rounded to.
- Underline the digit to the right of the circled digit.
- Decide whether the circled digit should stay the same or increase by 1.

If the underlined number is **less than** 5, the circled digit **stays the same**.
If the underlined number is **5 or greater**, the circled digit **increases by 1**.

- Write a zero or zeros as placeholders to the right of the digit that you circled.

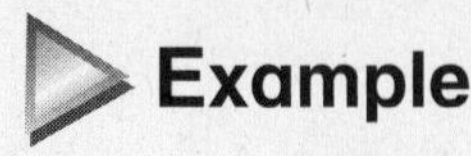

Round 698 to the nearest hundred.

The 6 is in the hundreds place, so circle it. The 9 is to the right of the 6, so underline it.

(6) 9 8

Since 9 is greater than 5, the 6 increases by 1 to 7. Write a 7 in the hundreds place, and write zeros as placeholders to the right of 7.

Therefore, 698 rounded to the nearest hundred is 700.

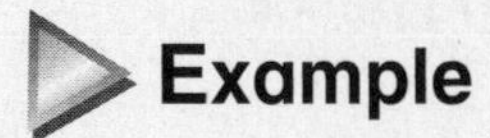

Round 32,153 to the nearest thousand.

The 2 is in the thousands place, so circle it. The 1 is to the right of 2, so underline it.

3 (2),1 53

Because 1 is less than 5, the 2 stays the same. Write zeros as placeholders to the right of 2.

Therefore, 32,153 rounded to the nearest thousand is 32,000.

Example

Round 512,384 to the nearest ten thousand.

The 1 is in the ten thousands place, so circle it. The 2 is to the right of 1, so underline it.

5 (1) 2, 3 8 4

Because 2 is less than 5, the 1 stays the same. Write zeros as placeholders to the right of 1.

Therefore, 512,384 rounded to the nearest ten thousand is 510,000.

When you round numbers to their greatest place, you can use place value and basic facts to compute sums and differences mentally.

Example

Estimate the sum of 6329 and 5482 without using paper and pencil.

6329 rounded to the nearest thousand is 6000.
5482 rounded to the nearest thousand is 5000.
Since $6 + 5 = 11$, the sum will be about 11 thousands, or 11,000.

Since both addends have been rounded down, the actual sum of 6329 and 5482 will be greater than 11,000.

Example

Estimate the sum of 75,802 and 49,783 without using paper and pencil.

75,802 rounded to the nearest ten thousand is 80,000.
49,783 rounded to the nearest ten thousand is 50,000.
Since $8 + 5 = 13$, the sum will be about 13 ten thousands, or 130,000.

Since both addends have been rounded up, the actual sum of 75,802 and 49,783 will be less than 130,000.

Sometimes rounding to the greatest place may not give an estimate as close as you would like. You can round one or both numbers to any place.

Example

Estimate the difference of 34,897 and 25,479.

34,897 rounded to the nearest ten thousand is 30,000.
25,479 rounded to the nearest ten thousand is 30,000.
The estimated difference is 0.

34,897 rounded to the nearest thousand is 35,000.
25,479 rounded to the nearest thousand is 25,000.
Now the estimated difference is 10,000, which is a more accurate estimate.

Example

Estimate the sum of 305,987 and 408,000.

Round only one number, 305,987, to the nearest thousand: 306,000.
306,000 + 408,000 = 300,000 + 400,000 + 6000 + 8000 = 714,000

Therefore, the estimated sum is 714,000.

Sometimes you can use the strategy of **grouping numbers** to estimate sums. (You cannot use this strategy to find differences.) In addition problems, the numbers can be switched around and grouped together so that you can more easily estimate sums.

Example

Estimate: 154 + 28 + 271 + 49

Switch 28 and 49 and group the first two numbers and the last two numbers. The sum of the first group is about 200. The sum of the second group is about 300.

(154 + 49) + (271 + 28)

200 + 300 = 500

Using grouping, the sum of the numbers is about 500.

Practice

1. Round 4766 to the nearest: ten __________ thousand __________

2. Round 28,035 to the nearest: thousand __________ ten thousand __________

3. Round 64,992 to the nearest: hundred __________ thousand __________

4. Round 50,871 to the nearest: ten __________ hundred __________

5. Round 721,534 to the nearest: hundred __________ thousand __________

6. Round 125,834 to the nearest: thousand __________ ten thousand __________

7. Round 386,917 to the nearest: hundred thousand __________

Directions: Estimate each sum or difference. Round each number to its greatest place.

8. 17,892 + 45,186

9. 35,213 − 24,907

10. 75,892 + 6090

11. 829,999 − 48,205

12. Explain how you could use grouping of the addends to estimate the sum for the following: 129 + 286 + 274

____

13. Jill wanted to estimate the sum of two numbers. After rounding each number to the nearest hundred, she used the expression 300 + 1800 to estimate the sum.

What are two numbers that Jill could have started with? Explain your answer.

____

____

CCSS: 4.OA.3, 4.NBT.4

Lesson 4: Adding Whole Numbers

When you join two amounts, you add. The answer to an addition problem is the **sum**. The numbers that are added are **addends**. To find the sum, line up the places and add each column, starting at the right. Remember to regroup when the sum in any column is greater than 9.

Example

Find the sum of 34,417 and 28,256.

regroup 10 thousands as 1 ten thousand → 1 1 **← regroup 10 ones as 1 ten**

$$\begin{array}{r} \scriptstyle 1\ \ \ \ 1 \\ 34{,}417 \\ +\ 28{,}256 \\ \hline 62{,}673 \end{array}$$

The sum of 34,417 and 28,256 is 62,673.

Use estimation to check the reasonableness of your answer. Rounded to the nearest thousand, 34,417 is about 34,000, and 28,256 is about 28,000. Since 34,000 + 28,000 = 62,000, the actual answer 62,673 is reasonable.

Example

Find the sum of 5182 and 94.

1 **← regroup 10 tens as 1 hundred**

$$\begin{array}{r} \scriptstyle 1\ \ \ \ \ \ \\ 5182 \\ +\ \ \ 94 \\ \hline 5276 \end{array}$$

The sum of 5182 and 94 is 5276.

Example

Use the information in the table to find the total number of acres of Farm I and Farm II.

Farm	Area in Acres
I	305
II	718

regroup 10 hundreds as 1 thousand → 1 1 **← regroup 10 ones as 1 ten**

$$\begin{array}{r} \scriptstyle 1\ \ 1\ \ \ \ \\ 305 \\ +\ 718 \\ \hline 1023 \end{array}$$

The total number of acres of Farm I and Farm II is 1023.

Addition Properties

The following **addition properties** are statements that are true for all whole numbers. The letters *a*, *b*, and *c* are **variables**. A variable is a letter or other symbol that can be used to represent any number.

Commutative Property of Addition The **order** of the numbers does not change the sum of an addition problem.
$a + b = b + a$ $2 + 4 = 4 + 2$ $6 = 6$
Associative Property of Addition The **grouping** of the numbers does not change the sum of an addition problem.
$(a + b) + c = a + (b + c)$ $(1 + 3) + 7 = 1 + (3 + 7)$ $4 \quad + 7 = 1 + 10$ $11 = 11$
Identity Property of Addition The sum of zero and any number is that number.
$3 + 0 = 3$

Understanding addition properties will help you with computation and problem solving.

Example

Use the associative property of addition to make this computation easier to do mentally: $22 + (18 + 34)$

Regroup the addends so you can make a ten.

$$\begin{aligned} 22 + (18 + 34) &= (22 + 18) + 34 \\ &= 40 + 34 \\ &= 74 \end{aligned}$$

Example

Use the commutative property of addition to make this computation easier to do mentally: $16 + 29 + 84$

Change the order of the addends so you can make a hundred.

$$\begin{aligned} 16 + 29 + 84 &= 16 + 84 + 29 \\ &= 100 + 29 \\ &= 129 \end{aligned}$$

Example

What number belongs in the blank to make the number sentence true?

$$34 + 50 = 50 + \underline{\quad ? \quad}$$

According to the commutative property of addition, the order does not change the sum of an addition problem. So, 34 + 50 is the same as 50 + 34.

The number 34 belongs in the blank to make the number sentence true.

Example

What number belongs in the blank to make the number sentence true?

$$(249 + 136) + 502 = 249 + (136 + \underline{\quad ? \quad})$$

According to the associative property of addition, the grouping does not change the sum of an addition problem. So, (249 + 136) + 502 is the same as 249 + (136 + 502)

The number 502 belongs in the blank to make the number sentence true.

Example

What number belongs in the blank to make the number sentence true?

$$\underline{\quad ? \quad} + 0 = 29{,}875$$

According to the identity property of addition, the sum of zero and any number is that number. So, 29,875 + 0 = 29,875

The number 29,875 belongs in the blank to make the number sentence true.

Sometimes you can use **compensation** to calculate sums mentally. Change one addend to the nearest 10 or 100 and then adjust the other addend to keep the balance.

Example

$$\begin{aligned} 48 + 34 &= (48 + 2) + (34 - 2) \\ &= 50 + 32 \\ &= 82 \end{aligned}$$

An **expression** represents a mathematical value. It can be a number or more than one number along with one or more operation signs. An **equation** is a number sentence stating that two expressions are equal.

You can use an equation to represent the information in a word problem.
Use a symbol such as the letter *n* or □ to represent the unknown number.
Then decide on an operation, write the equation, and use it to solve the problem.

Example

In June, 34,968 people visited the Statue of Liberty. In July, 39,051 people visited the Statue of Liberty. What is the total number of visitors in June and July?

The problem is asking that you join the two groups of visitors to find the total for June and July. When two amounts are being joined, use the addition operation.

Write an addition equation that represents the information in the problem.
34,968 + 39,051 = total number of visitors in June and July
34,968 + 39,051 = □

To solve the equation, write the addends in vertical form.
Line up digits with the same place values. Find the sum.

$$\begin{array}{r} 34{,}968 \\ +\ 39{,}051 \\ \hline 74{,}019 \end{array}$$

34,968 + 39,051 = 74,019

A total of 74,019 people visited the Statue of Liberty in June and July.

To check that your answer is reasonable, estimate.

Rounded to the nearest thousand, 34,968 is about 35,000 and 39,051 is about 39,000. The total is 74,000, so the answer is reasonable.

Example

In August, 21,772 people visited the Alamo. In September, there were 824 more visitors to the Alamo than there were in August. What is the total number of visitors to the Alamo in August and September?

The problem is asking you to join groups of visitors to find the total number of visitors, so you will use addition. Notice that the problem does not tell you the number of visitors there were in September. So the first step is to write and solve an equation to find the number of September visitors.

$21{,}772 + 824 = s$ **The letter *s* represents the number of September visitors.**

To solve the equation, write the addends in vertical form.
Line up digits with the same place values. Find the sum.

$$\begin{array}{r} 21{,}772 \\ +\quad 824 \\ \hline 22{,}596 \end{array}$$

Solve the first equation: 21,772 + 824 = 22,596
So there were 22,596 visitors in September.

Then use your answer to write another equation to find the total number of visitors.

$21{,}772 + 22{,}596 = t$ **The letter *t* represents the total number of visitors.**

Write the answers in vertical form and line up digits with the same place values. Find the sum.

$$\begin{array}{r} 21{,}772 \\ +\ 22{,}596 \\ \hline 44{,}368 \end{array}$$

Solve the second equation: 21,772 + 22,596 = 44,368
So a total of 44,368 people visited the Alamo in August and September.

To check that your answer is reasonable, estimate.

For the first equation, the sum is about 22,000 + 1000, or 23,000.
For the second equation, the sum is about 22,000 + 23,000, or 45,000.
Since all the addends were rounded up to the greater thousand, the actual answer should be less than the estimate.

So the actual answer of 44,368 is reasonable.

Practice

Directions: For questions 1 through 10, line up the digits with the same place values. Then find the sum.

1. 579 + 724 = __________
2. 1539 + 48 = __________
3. 2856 + 4630 = __________
4. 429 + 12,708 = __________
5. 24,988 + 19,837 = __________
6. 52,911 + 2789 = __________
7. 39,083 + 715 = __________
8. 4819 + 6327 = __________
9. 30,793 + 62,684 = __________
10. 7002 + 12,956 = __________

11. LaToya is solving the addition problem below.

$$\begin{array}{r} 1514 \\ +\ 3865 \\ \hline 79 \end{array}$$

What should LaToya do next?

____

____

Directions: For questions 12 and 13, use addition to solve each problem.

12. On Saturday, 2563 people visited the Science Museum. On Sunday, 893 people visited the Science Museum. How many people visited the Science Museum on both days?

    A. 2456
    B. 3456
    C. 3466
    D. 11,493

13. In July, an amusement park collected $38,956 in entrance fees for adults and $10,051 in entrance fees for children. What is the total amount collected in entrance fees in July?

    A. $48,007
    B. $48,907
    C. $49,007
    D. $59,007

Solve each problem.

14. Denzel bought a car that cost $12,395 and a computer that costs $643. How much did Denzel spend altogether?

 $13038

15. Two candidates received votes in a local election. Miss Jackson received 17,394 votes, and Mr. King received 16,901 votes. How many votes were cast in all?

 34295 votes

16. A factory manufactured 9388 calculators in January and 10,719 calculators in February. How many calculators were manufactured in both months?

 1110 calculators

17. Bridgeville has a population of 45,752. Rockport has a population that is 1506 greater than the population of Bridgeville. What is the population of the two cities combined?

 Explain how you got your answer.

 I added 45752 + 1506 = 47258 then added 45752 + 47258 = 94010 votes

 45752 + 1506 = 47258
 47258 + 45752 = 94010

Lesson 5: Subtracting Whole Numbers

There are three parts to a subtraction problem.

The number you are subtracting from is the **minuend**. → 795
The number you are subtracting is the **subtrahend**. → – 382
The answer to a subtraction problem is the **difference**. → 413

When finding the difference, remember to "regroup" where you need to.

Example

Find the difference of 319 and 198.

```
 211   ← regroup 1 hundred as 10 tens
 319
– 198
 ----
 121
```

The difference of 319 and 198 is 121.

Example

Find the difference of 29,745 and 684.

```
      614  ← regroup 1 hundred as 10 tens
  29,745
 –   684
 -------
  29,061
```

The difference of 29,745 and 684 is 29,061.

Example

There were 35,634 fans at the beginning of a football game. By the end of the game there were 17,022 fans. How many fans left the game before it was over?

```
  215     ← regroup 1 ten thousand as 10 thousands
  35,634
– 17,022
 -------
  18,612
```

There were 18,612 fans that left the game.

TIP: Because addition and subtraction are "opposites," you can use addition to check your answer to a subtraction problem. When you add the difference and the number that you subtracted, your sum should be the number that you subtracted from.

CCSS: 4.OA.3, 4.NBT.4

Sometimes you can use **compensation** to find a difference mentally. When you use compensation to subtract, you have to do the same thing to each number. Try to change the subtrahend (second number) to the nearest 10 or 100.

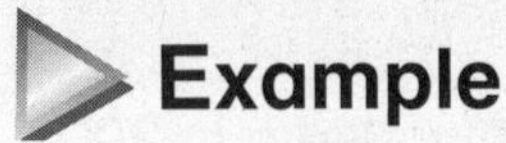

Example

$$\begin{aligned} 844 - 205 &= (844 - 5) - (205 - 5) \\ &= 839 - 200 \\ &= 639 \end{aligned}$$

You can use an equation to represent the information in a subtraction word problem. Use a variable such as *n* or a symbol such as □ or ○ to represent the unknown number. Then you can use subtraction to solve the problem.

Example

In April, 56,238 people visited Washington, D.C. If 31,062 of the visitors were adults, how many of the visitors were children?

The problem is asking how many children visited Washington, D. C., in April. The given information is the total number of visitors and the number of visitors who were adults. To find the number of visitors who were children, you find the difference for the total number of visitors and the number of adults. To find the difference, you subtract.

Write a subtraction equation that represents the information in the problem.
56,238 − 31,062 = number of visitors who were children
56,238 − 31,062 = □

To solve the equation, write the minuend and subtrahend in vertical form. Line up the digits with the same place value. Find the difference.

$$\begin{array}{r} 56{,}238 \\ -\ 31{,}062 \\ \hline 25{,}176 \end{array}$$

56,238 − 31,062 = 25,176

Exactly 25,176 children visited Washington, D. C., in April.

Estimate to check the reasonableness of your answer.

Rounded to the nearest thousand, 56,238 is about 56,000 and 31,062 is about 31,000. The difference is 25,000, so the answer is reasonable.

Example

In 2008, 53,226 people graduated from college. In 2009, 40,773 men and 29,531 women graduated from college. How many more graduates were there in 2009 than in 2008?

The problem is asking how many more people graduated from college in 2009 than in 2008, so you will subtract the 2008 number of graduates from the 2009 number. However, the total number of 2009 graduates is not given. So the first step is to write and solve an addition equation to find the total number of 2009 graduates.

$40{,}773 + 29{,}531 = g$ **The letter *g* represents the total number of 2009 graduates.**

To solve the equation, write the addends in vertical form.
Line up digits with the same place values. Find the sum.

$$\begin{array}{r} 40{,}773 \\ +\ 29{,}531 \\ \hline 70{,}304 \end{array}$$

Solve the first equation: $40{,}773 + 29{,}531 = 70{,}304$
So there were 70,304 graduates in 2009.

Then write an equation to find how many more people graduated in 2009 than in 2008.

$70{,}304 - 53{,}226 = m$ **The letter *m* represents how many more graduates there were in 2009.**

To solve the equation, write the minuend and subtrahend in vertical form.
Line up digits with the same place values. Find the difference.

$$\begin{array}{r} 70{,}304 \\ -\ 53{,}226 \\ \hline 17{,}078 \end{array}$$

Solve the second equation: $70{,}304 - 53{,}226 = 17{,}078$

There were 17,078 more college graduates in 2009 than in 2008.

To check that your answer is reasonable, estimate.

For the first equation, the sum is about $40{,}000 + 30{,}000$, or 70,000.

For the second equation, the difference is about $70{,}000 - 53{,}000$, or 17,000.

So the actual answer of 17,078 is reasonable.

CCSS: 4.OA.3, 4.NBT.4

Practice

Directions: For questions 1 through 10, line up the digits with the same place values. Then find the difference.

1. 925 − 682 = ______________
2. 3806 − 71 = ______________
3. 8136 − 7511 = ______________
4. 25,809 − 976 = ______________
5. 44,160 − 26,858 = ______________
6. 73,612 − 4935 = ______________
7. 67,608 − 249 = ______________
8. 8392 − 2750 = ______________
9. 51,497 − 19,622 = ______________
10. 36,388 − 2779 = ______________

11. Lauren is solving the subtraction problem below.

```
   512
  5462
− 2178
  ----
     4
```

What should Lauren do next?

____

____

Directions: For questions 12 and 13, use subtraction to solve each problem.

12. In June, 7122 people visited the Washington Monument and 6013 people visited the Lincoln Memorial. How many more people visited the Washington Monument than the Lincoln Memorial?

    A. 1109

    B. 1111

    C. 1119

    D. 2109

13. In October, people at a movie theater spent $62,550 on tickets and $41,355 on snacks. How much less was spent on snacks than on tickets?

    A. $20,095

    B. $20,195

    C. $21,195

    D. $21,205

Directions: For questions 14 and 15, use compensation to calculate the answer mentally.

14. 499 + 175 = ?

A. 674 C. 684

B. 676 D. 774

15. 821 − 202 = ?

A. 602 C. 619

B. 617 D. 623

Solve each problem.

16. A garden nursery planted 562 tulip bulbs and 713 daffodil bulbs. How many more daffodil bulbs than tulip bulbs did the nursery plant?

____

17. Last year, Al's Autos sold 8074 vehicles. If 3149 of the vehicles sold were trucks, how many of the vehicles sold were **not** trucks?

____

18. There were 16,405 fans at a basketball game on Friday. There were 23,112 fans at a basketball game on Saturday. How many more fans attended the game on Saturday than attended the game on Friday?

____

19. Cooperstown has a population of 43,295. Yorkville has a population that is 4067 less than the population of Cooperstown. What is the population of the two cities combined?

Explain how you got your answer.

____

____

____

Unit 1 Practice Test

1. In the base-ten numeral 770, in which place is the digit that represents a value that is 10 times the value of the 7 in the tens place?

   hundreds

2. In the base-ten numeral 36,183, what digit is in the thousands place?

   6

3. What is 12,834 rounded to the nearest hundred?

4. Which of the following numbers is greater than 38,898?

   38,861; 38,957; 38,799; 38,895

   38,957

5. The population of Lakeford is 5736. What is the number name for 5736?

   Fifty seven thousand thirty six

6. In the number 288, explain how the value represented by the digit in the tens place is related to the value represented by the digit in the ones place.

   the 8 in the tens place and the ones plase are the same number

For questions 7-12, choose the correct answer.

7. Which is the base-ten numeral for 300,000 + 40,000 + 80 + 7?

   A. 348,007

   B. 340,807

   C. 340,087

   D. 304,087

8. Exactly 23,118 men and 26,053 women voted in an election. How many people voted in all?

   A. 49,161

   B. 49,171

   C. 49,271

   D. 50,171

9. In which base-ten numeral does the digit in the ten-thousands place represent a value that is 10 times the value represented by the digit in the thousands place?

   A. 38,552

   B. 41,335

   C. 66,093

   D. 84,281

10. What is the expanded form for 58,063?

    A. 5000 + 800 + 60 + 3

    B. 50,000 + 8000 + 60 + 3

    C. 50,000 + 800 + 60 + 3

    D. 50,000 + 80,000 + 6000 + 3

11. The students at Spring Lake Elementary collected 4327 cans for recycling. The students at Kennedy Street School collected 3819 cans for recycling. How many more cans did the students at Spring Lake Elementary collect than the students at Kennedy Street School?

    A. 502

    B. 507

    C. 508

    D. 1508

12. What is 7631 rounded to the nearest hundred?

    A. 7600

    B. 7650

    C. 7700

    D. 8000

Use the table to answer questions 13 through 17.

City Populations

City	Population
Upton Park	6082
Delmore	4177
Yorkville	6695
Lincoln City	3859
Franklin	6659

6082
+6695
12,777

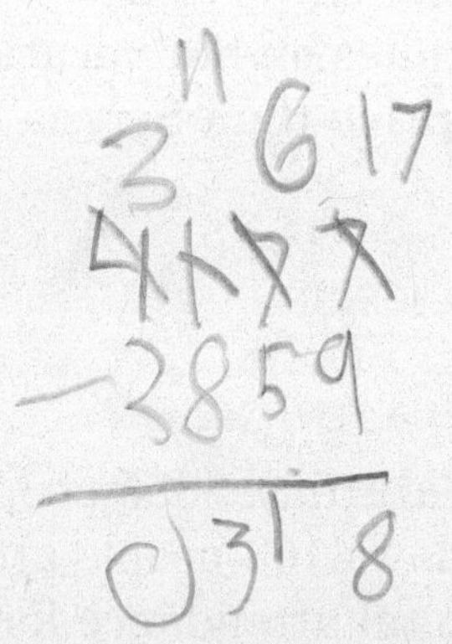

13. Which city has the greatest population?

Lincon city

14. Which city has a population greater than 6000 and less than 6500?

Upton Park

15. What is the population of Franklin rounded to the nearest ten?

6660

16. How many more people live in Delmore than in Lincoln City?

318 People

17. What is the combined population of Upton Park and Yorkville?

12,777

Use estimation to see if your answer is reasonable. Tell which estimation method you used, and explain why you chose that method.

I chose to round up to the nearest thousand so I can get a greater number

Solve each problem.

18. Ben drove 509 miles on his vacation. Serena drove 384 miles on her vacation. How many more miles did Ben drive than Serena?

    125 miles

19. On Saturday, 3855 people attended a tennis match. On Sunday, 4163 people attended a tennis match. What was the total attendance for Saturday and Sunday?

    8018 People

20. A factory manufactured 43,928 cars in May and 51,608 cars in June. How many cars were manufactured in both months?

    95,536 cars

21. Mr. Cohen earned $39,552 in 2009. Mr. King earned $605 more than Mr. Cohen in 2009.
    What are the combined earnings of both men in 2009?

    $79,709

    Explain how you got your answer.

    I added 39552 and 605 and I got 40157 and I added Both numbers 39552 and 40157

22. Mrs. Sanchez wrote the following numbers on the board.

1366 1404 1482 1319

Part A

Write each number in the place-value table below.

Hundred Thousands	Ten Thousands	Thousands	Hundreds	Tens	Ones
		1	3	6	6
		1	4	0	4
		1	4	8	2
		1	3	1	9

Part B

Which number is **greater than** 1400 but **less than** 1450?

1,404

Part C

Which number is **greatest** and which number is **least?**

greatest 1,482 least 1,319

Explain how you found your answer.

I found the ghearest I looked at the hundreds and I saw 4 and then I looked at the tens and I saw 1482 was the greatost

23. In a local school, there are 298 boys and 317 girls.

Part A
Estimate the total number of students in the school.

600

Explain how you got your answer.

I looked at the tens place and I couldent round so I looked at the other number and then I could round up

298
+317
615

Part B
How many students are in the school?

615 students

Explain how you got your answer.

I added 198 and 317 and I got 615 students

Part C
Is your exact answer reasonable? Explain your thinking.

my awnser is reasonagal becaase when I rounded I got 600 and it is close to 615.

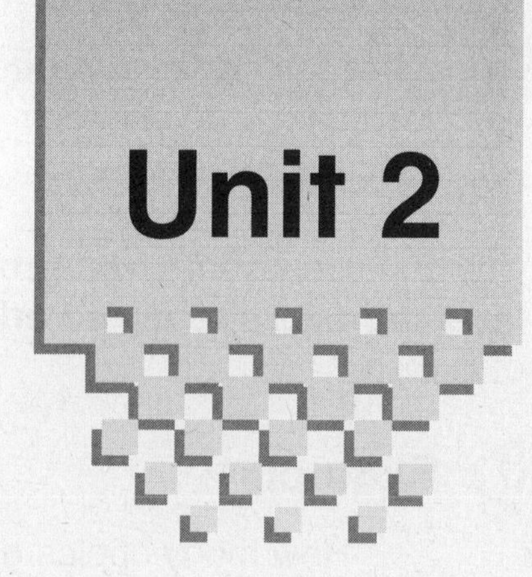

Operations and Algebraic Thinking

Anytime you compute the answer to a problem, you use at least one of the four mathematical operations—addition, subtraction, multiplication, and division. You use these operations to solve problems that come up in everyday situations—real-world problems that your knowledge of math can help you solve.

In this unit, you will review multiplication and division facts and strategies. You will use place value and properties as you multiply and divide whole numbers. You will estimate and learn some strategies for mentally computing products and quotients. You will learn about factors and multiples. You will analyze different types of patterns. Throughout the unit, you will use equations with variables to solve multiplication and division problems.

In This Unit

Lesson 6: Multiplication Facts

The answer to a multiplication problem is a **product**. The numbers you multiply are **factors**. Here is one way to write a multiplication fact:

$3 \times 4 = 12 \leftarrow$ **product**

↑ ↑

factors

You can use **equal-sized groups** to multiply. After you group items together, you can **skip count**, use **repeated addition**, or use an **equation** to find the total number of items.

Example

How many apples are there? Separate the apples into groups of 4.

You can use skip counting to find the total number of apples.

4, 8, 12

You can use a repeated addition sentence to show the number of apples.

$4 + 4 + 4 = 12$

You can use an equation to show the number of apples.

$3 \times 4 = 12$

There are 12 apples.

Both addition and multiplication involve joining groups, but multiplication makes it easier to join groups that have the same number of objects.

Example

This is a group of 5 circles.

○○○○○

This group has **5 more** circles than the first group.

$5 + 5 = 10$

$2 \times 5 = 10$

○○○○○
○○○○○

This group has **5 times as many** circles as the first group.

$5 + 5 + 5 + 5 + 5 = 25$

$5 \times 5 = 25$

○○○○○
○○○○○
○○○○○
○○○○○
○○○○○

CCSS: 4.OA.1, 4.OA.2, 4.NBT.5

Another way to show multiplication is to use arrays. An **array** is a group of objects that are arranged in rows and columns in the shape of a rectangle.

Example

How many pencils are there?

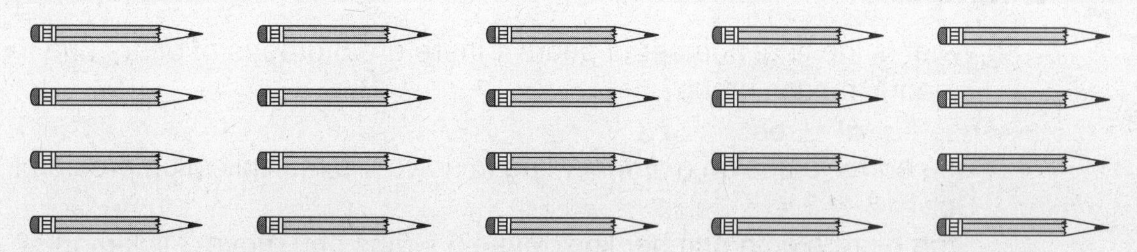

One way to find the answer is to count the pencils, but this can take a long time. Instead, you can multiply the number of columns by the number of rows.

There are 4 rows of pencils, so 4 is the first factor.
There are 5 pencils in every row, so 5 is the second factor.

There are $4 \times 5 = 20$ pencils. You can check by counting the pencils.

The model shows that 20 pencils is 4 times as many pencils as 5 pencils. You can turn the model to show 5 rows of 4 pencils. The total number of pencils is 20, and now the model shows that 20 pencils is 5 times as many pencils as 4 pencils.

So 4×5 pencils $= 20$ pencils and 5×4 pencils $= 20$ pencils.

You can also show multiplication using an area model. An **area model** is a rectangle made of squares that are each 1 unit by 1 unit.

Example

What multiplication problem is shown by the area model?

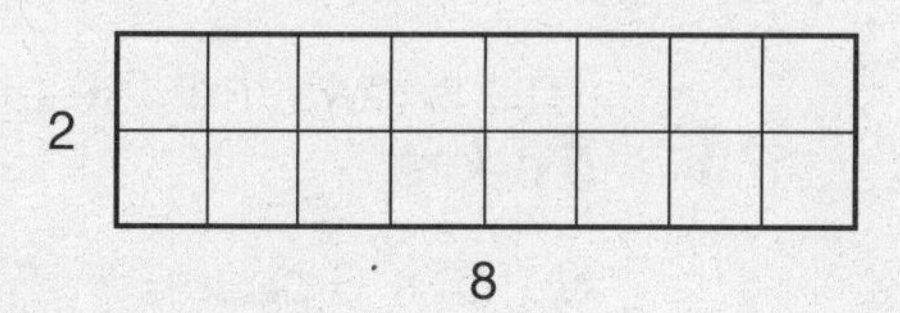

There are 2 rows of squares, so 2 is the first factor.
There are 8 squares in each row, so 8 is the second factor.
The area model shows $2 \times 8 = 16$. You can check by counting the squares.

The model shows that 16 squares is 2 times as many squares as 8 squares. You can turn the model to show 8 rows of 2 squares each. The total number of squares is 16, and now the model shows that 16 squares is 8 times as many squares as 2 squares.

So 2×8 squares $= 16$ squares, and 8×2 squares $= 16$ squares.

You can solve a multiplication problem by making equal-sized jumps on a **number line.** The number of jumps you make is one of the factors.
The number of tick marks in each jump is the other factor. The number of tick marks you move in all of the jumps is the product.

Example

What is the total number of pears if there are 5 groups of pears and there are 3 pears in each group?

Use equal jumps on a number line to solve the multiplication problem.
Solve: $5 \times 3 = ?$
Start at zero on a number line. Make 5 jumps and move 3 tick marks in each jump.

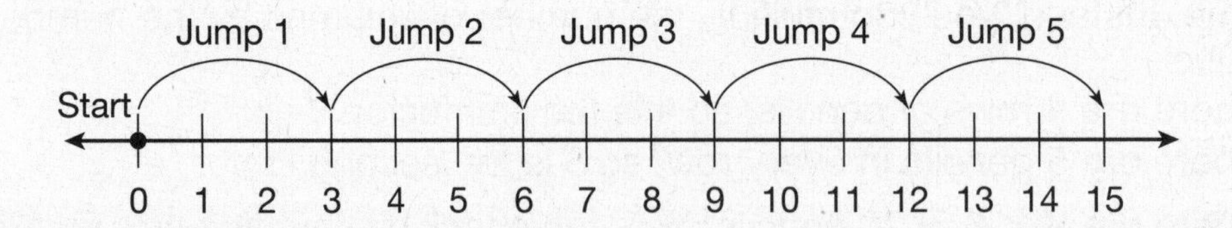

The number where you stop shows the total number of tick marks you have moved. This is the same as the total number of pears. The number line shows that 15 is the product, and that 15 pears is 5 times as many as 3 pears.

If there were 3 groups of 5 pears, you would find the same product by making 3 jumps and moving 5 tick marks in each jump.

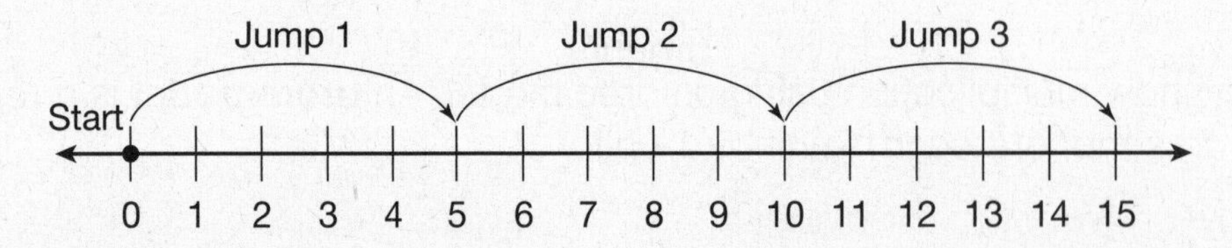

This shows that the product is 15, and that 15 pears is 3 times as many as 5 pears.

So, 5×3 pears $=$ 15 pears, and 3×5 pears $=$ 15 pears.

CCSS: 4.OA.1, 4.OA.2, 4.NBT.5

Practice

Directions: Use the following drawing to answer questions 1 and 2.

1. Use skip counting by 2s to show how many number cubes there are.

   ________, ________, ________, ________, ________, ________

2. Write a repeated addition sentence to show how many number cubes there are.

   ________ + ________ + ________ + ________ + ________ + ________ = ________

Directions: In questions 3 and 4, find the factors for each array or area model. Then, find the product.

3.

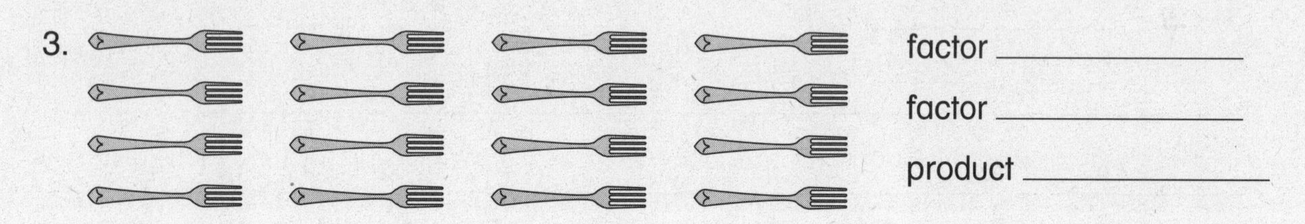

factor ________________

factor ________________

product ________________

4.

factor ________________

factor ________________

product ________________

5. There are 2 boxes with 5 crayons in each box. How many crayons are there in all?

   A. 7

   B. 8

   C. 10

   D. 15

6. There are 8 rows of triangles with 4 triangles in each row. How many triangles are there?

   A. 12

   B. 16

   C. 30

   D. 32

7. Use the number line to multiply 2×7.

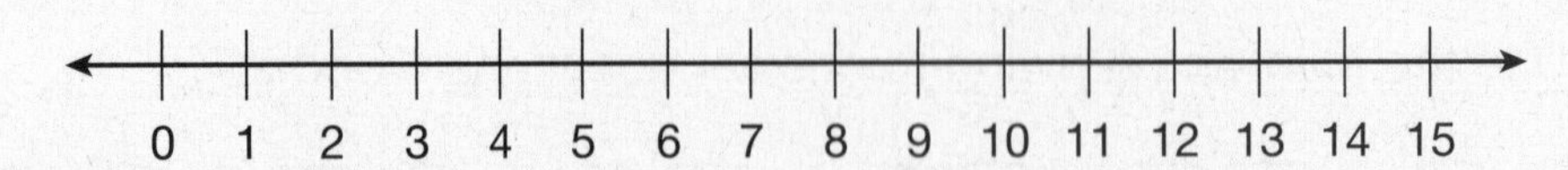

$2 \times 7 =$ ______________

Directions: For questions 8 through 13, calculate each product. Explain the strategy that you used.

8. $6 \times 3 =$ ______________

____

____

9. $5 \times 5 =$ ______________

____

____

10. $9 \times 7 =$ ______________

____

____

11. $8 \times 6 =$ ______________

____

____

12. $5 \times 10 =$ ______________

____

____

13. $8 \times 12 =$ ______________

____

____

CCSS: 4.OA.1, 4.OA.2, 4.NBT.5

14. Fill in the multiplication table. Multiply a number in the left column times a number in the top row. Write the product in the box where the column and row meet.

×	1	2	3	4	5	6	7	8	9	10	11	12
1												
2												
3												
4												
5												
6												
7												
8												
9												
10												
11												
12												

15. Daniel and Donna collect small glass figures. Daniel has 6 times as many figures as Donna has. Donna has 3 small figures. Write and solve an equation to show how many figures are in Daniel's collection.

    Use *d* for the number of figures in Daniel's collection.

    ____

Lesson 7: Multiplying with Multiples of 10, 100, and 1000

A **multiple** of a whole number is found by multiplying that number by any other whole number. Multiples of 10, such as 10, 20, 30, 40, and so on, have one zero at the end. Multiples of 100, such as 100, 200, 300, 400, and so on, have two zeros at the end,. Multiples of 1000, such as 1000, 2000, 3000, 4000, and so on, have three zeros at the end. A pattern is formed when a number is multiplied by 10, 100, or 1000.

Example

$1 \times \mathbf{10} = \mathbf{10}$ $\quad 1 \times \mathbf{100} = \mathbf{100}$ $\quad 1 \times \mathbf{1000} = \mathbf{1000}$

$2 \times \mathbf{10} = \mathbf{20}$ $\quad 2 \times \mathbf{100} = \mathbf{200}$ $\quad 2 \times \mathbf{1000} = \mathbf{2000}$

$3 \times \mathbf{10} = \mathbf{30}$ $\quad 3 \times \mathbf{100} = \mathbf{300}$ $\quad 3 \times \mathbf{1000} = \mathbf{3000}$

$4 \times \mathbf{10} = \mathbf{40}$ $\quad 4 \times \mathbf{100} = \mathbf{400}$ $\quad 4 \times \mathbf{1000} = \mathbf{4000}$

Notice the pattern. When you multiply 10 and any 1-digit number, there is always a zero in the ones place of the product and the number in the tens place is the same as the 1-digit number. A similar pattern is formed when a number is multiplied by a multiple of 10, 100, or 1000. As the number of zeros increases in the factors, the number of zeros increases by that same amount in the product. Using that pattern along with a basic multiplication fact, you can find the products without using pencil and paper.

Example

Use the fact to find the products: $3 \times 6 = 18$

$3 \times 60 = ?$ $\quad 3 \times 6 = 18$, so $3 \times \mathbf{60} = 18\mathbf{0}$

$3 \times 600 = ?$ $\quad 3 \times 6 = 18$, so $3 \times \mathbf{600} = 18\mathbf{00}$

$3 \times 6000 = ?$ $\quad 3 \times 6 = 18$, so $3 \times \mathbf{6000} = 18,\mathbf{000}$

Notice that the pattern is the same as for multiplying by 10, 100, and 1000. Each time you increase the number of zeros in the factors, the number of zeros in the product increases by the same amount.

You need to be especially careful writing zeros in the product when the fact you are using already has a zero in the product.

Example

Use the fact to find the products: $8 \times 5 = 40$

$8 \times 50 = ?$

$8 \times 500 = ?$

$8 \times 5000 = ?$

Think about the place-value names:

8×5 ones = 40 ones, or 40

8×5 tens = 40 tens, or 4**0**0, so 8×5**0** = 4**0**0

8×5 hundreds = 40 hundreds, or 40**00**, so 8×5**00** = 40**00**

8×5 thousands = 40 thousands, or 40,**000**, so 8×5**000** = 40,**000**

You can also use this rule to multiply by a multiples of 10,100 or 1000: **Remove any zeros from the factors, and multiply just the nonzero digits. Then put the zeros back onto the product.**

Examples

3 × 6**0** → 3 × 6 = 18 → 18**0** (Replace the zero.)

4**0** × 2**0** → 4 × 2 = 8 → 8**00** (Replace the two zeros.)

2**00** × 11 → 2 × 11 = 22 → 22**00** (Replace the two zeros.)

5 × 12**0** → 5 × 12 = 60 → 60**0** (Replace the zero.)

Practice

Directions: For questions 1 and 2, use multiplication facts and patterns to solve the problems.

1. $10 \times 100 = ?$

   A 10
   B 100
   C 1000
   D 10,000

2. $5 \times 60 = ?$

   A 30
   B 300
   C 3000
   D 30,000

Directions: For questions 3 through 22, find the product.

3. $3 \times 40 =$ ______________

4. $9000 \times 6 =$ ______________

5. $4 \times 50 =$ ______________

6. $400 \times 8 =$ ______________

7. $50 \times 50 =$ ______________

8. $70 \times 30 =$ ______________

9. $8 \times 100 =$ ______________

10. $1100 \times 3 =$ ______________

11. $80 \times 50 =$ ______________

12. $7 \times 7000 =$ ______________

13. $10 \times 4 =$ ______________

14. $300 \times 60 =$ ______________

15. $100 \times 500 =$ ______________

16. $500 \times 7 =$ ______________

17. $9 \times 90 =$ ______________

18. $20 \times 700 =$ ______________

19. $8 \times 2000 =$ ______________

20. $900 \times 6 =$ ______________

21. $100 \times 100 =$ ______________

22. $4000 \times 2 =$ ______________

Directions: For questions 23 and 24, use multiplication facts to solve the problems.

23. $3 \times 200 = ?$

    A. 6
    B. 60
    C. 600
    D. 6000

24. $400 \times 12 = ?$

    A. 48
    B. 480
    C. 4800
    D. 48,000

Directions: For questions 25 through 30, multiply using the rule shown on page 47. Show your work in the space below each problem.

25. $1200 \times 3 =$ ______________

26. $7 \times 90 =$ ______________

27. $30 \times 60 =$ ______________

28. $8 \times 7000 =$ ______________

29. $500 \times 40 =$ ______________

30. $20 \times 20 =$ ______________

31. $40 \times 600 =$ ______________

    Explain how you got your answer.

    ____

    ____

32. $7 \times 9000 =$ ______________

    Explain how you got your answer.

    ____

    ____

Lesson 8: Multiplication Properties

The following **multiplication properties** are statements that are true for all whole numbers. The letters *a*, *b*, and *c* are **variables**. A variable is a letter or other symbol that can be used to represent any number.

Associative Property of Multiplication The **grouping** of the numbers does not change the product of a multiplication problem.
$(a \times b) \times c = a \times (b \times c)$ $(3 \times 2) \times 5 = 3 \times (2 \times 5)$ $6 \times 5 = 3 \times 10$ $30 = 30$
Commutative Property of Multiplication The **order** of the numbers does not change the product of a multiplication problem.
$a \times b = b \times a$ $3 \times 8 = 8 \times 3$ $24 = 24$
Identity Property of Multiplication The product of any number and 1 is that number.
$a \times 1 = a$ $4 \times 1 = 4$ $4 = 4$
Zero Property of Multiplication The product of any number and 0 is 0.
$a \times 0 = 0$ $5 \times 0 = 0$ $0 = 0$

CCSS: 4.NBT.5

Understanding multiplication properties will help you with computation and problem solving. Using the properties will simplify the computations.

Example

Use the **associative property of multiplication** to simplify the problem.

$20 \times (10 \times 12) = ?$

According to the associative property of multiplication, the grouping of the numbers does not change the product.

So, regroup the multiplication problem so that the simpler multiplication is done first.

$(20 \times 10) \times 12$

Then multiply by 12.

$$(20 \times 10) \times 12 = 200 \times 12$$
$$= 2400$$

Example

Use the **commutative property of multiplication** to find the number that belongs in the blank to make the number sentence true.

$8 \times 10 = 10 \times$ ___?___

According to the commutative property of multiplication, the order does not change the product.

So, 8×10 is the same as 10×8. The number 8 belongs in the blank to make the number sentence true.

$8 \times 10 = 10 \times 8$

Example

Use the **identity property of multiplication** to find the answer without pencil and paper.

$394 \times 1 = ?$

According to the identity property, the product of any number and 1 is that number.

So, $394 \times 1 = 394$

Example

Use the **zero property of multiplication** to find the answer without pencil and paper.

$273 \times 0 = ?$

According to the zero property of multiplication, the product of any number and 0 is 0.

So, $273 \times 0 = 0$

Practice

Directions: For questions 1 through 6, use number properties to fill in the missing numbers.

1. $4615 \times 1 = 1 \times$ ________

2. $10 \times (7 \times 3) = ($________ $\times$ ________$) \times 3$

3. $23 \times 5 = 5 \times$ ________

4. $85 \times 216 = 216 \times$ ________

5. $365 \times 0 =$ ________

6. $3 \times$ ________ $= 3$

7. Show that $5 \times 9 = 9 \times 5$.

8. Show that $2 \times (4 \times 3) = (2 \times 4) \times 3$.

9. Which is the same as $2 \times (5 \times 9)$?

   A. $(2 \times 5) \times 9$

   B. $(2 \times 5) \times (2 \times 9)$

   C. $(2 \times 5) + 9$

   D. $2 \times (5 + 9)$

10. Which is the same as 12×3?

   A. $12 \times (3 \times 12)$

   B. 3×12

   C. 1×12

   D. $3 + 12$

11. $65 \times 1 = ?$

   A. 0

   B. 64

   C. 65

   D. 66

12. $294 \times 0 = ?$

   A. 0

   B. 1

   C. 294

   D. 2940

Directions: For questions 13 through 22, write the number property used.

13. $(3 \times 2) \times 3 = 3 \times (2 \times 3)$ ____________________

14. $18 \times 9 = 9 \times 18$ ____________________

15. $32 \times 0 = 0$ ____________________

16. $5 \times (4 \times 2) = (5 \times 4) \times 2$ ____________________

17. $8 \times 75 = 75 \times 8$ ____________________

18. $487 \times 1 = 487$ ____________________

19. $16 \times (4 \times 9) = (16 \times 4) \times 9$ ____________________

20. $52 \times 3 = 3 \times 52$ ____________________

21. $4967 \times 0 = 0$ ____________________

22. $(7 \times 8) \times 3 = 7 \times (8 \times 3)$ ____________________

Directions: For questions 23 and 24, calculate each answer. Name the property you used and explain how you used it.

23. $(9 \times 30) \times 10 =$

____

____

24. $483 \times 1 =$

____

____

CCSS: 4.NBT.5

Lesson 9: The Distributive Property

You can use the **distributive property** to rename one factor as a sum. This can make it easier for you to use mental computation to find a sum.

Distributive Property This property relates the operations of addition and multiplication.
$a \times (b + c) = (a \times b) + (a \times c)$
$2 \times (7 + 4) = (2 \times 7) + (2 \times 4)$
$2 \times 11 = 14 + 8$
$22 = 22$

You can use an array to show the distributive property.

Example

Use a 4×9 array to show that $4 \times (6 + 3) = (4 \times 6) + (4 \times 3)$.

XXXXXXXXX

XXXXXXXXX

XXXXXXXXX

XXXXXXXXX

Separate the 4×9 array into one 4×6 array and one 4×3 array. Since the total number of Xs has not changed, you can see that 4×9 represents the same number as $(4 \times 6) + (4 \times 3)$.

XXXXXX XXX

XXXXXX XXX

XXXXXX XXX

XXXXXX XXX

So $4 \times (6 + 3) = (4 \times 6) + (4 \times 3)$.

You can the distributive property to help you multiply greater numbers.

Example

$9 \times 34 = ?$

Step 1: Rename one factor as a sum.
In this case, it helps to rename 34 using the expanded form $30 + 4$. So rewrite the problem as $9 \times (30 + 4)$.

Step 2: Multiply each addend by 9.
So $9 \times (30 + 4) = (9 \times 30) + (9 \times 4)$.

Step 3: Use facts to help you solve the problem mentally. Add the products.
$9 \times 3 = 27$, so $9 \times 30 = 270$ and $9 \times 4 = 36$.

Since $270 + 36 = 306$, $9 \times 34 = 306$.

Example

$3 \times 27 = ?$

Use expanded notation to rename 27 as $20 + 7$. Use place-value models of tens and ones to show the problem as $3 \times (20 + 7)$. Then multiply each addend by 3.

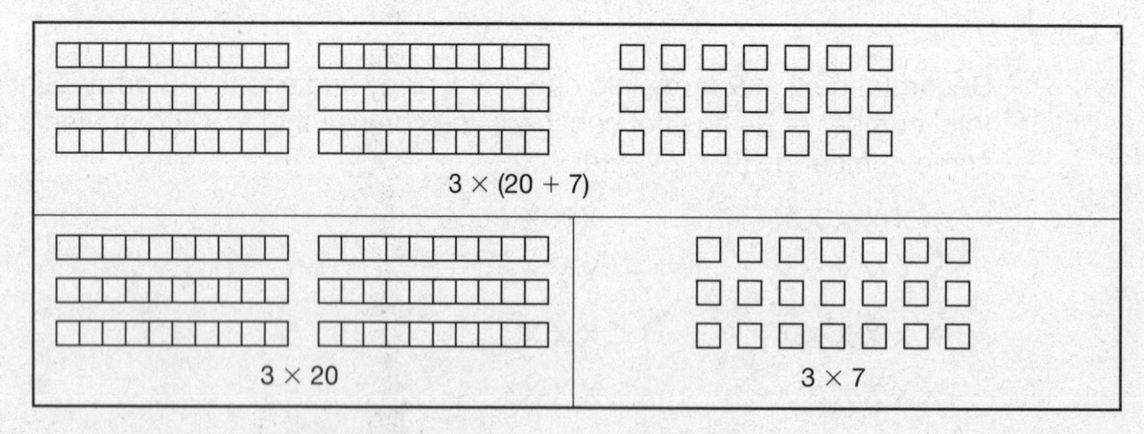

$3 \times 20 = 60$ and $3 \times 7 = 21$.

Since $60 + 21 = 81$, $3 \times 27 = 81$

CCSS: 4.NBT.5

Example

$20 \times 45 = ?$

Step 1: Use expanded notation to rename one factor as a sum.
$20 \times (40 + 5)$

Step 2: Multiply each addend by 20.
$(20 \times 40) + (20 \times 5) = 800 + 100$

Step 3: Add the products.
$800 + 100 = 900$

So $20 \times 45 = 900$.

Example

$8 \times 16 = ?$

Use expanded notation to rename 16 as 10 + 6. Use an area model to show the problem as $8 \times (10 + 6)$. Multiply each addend by 8.

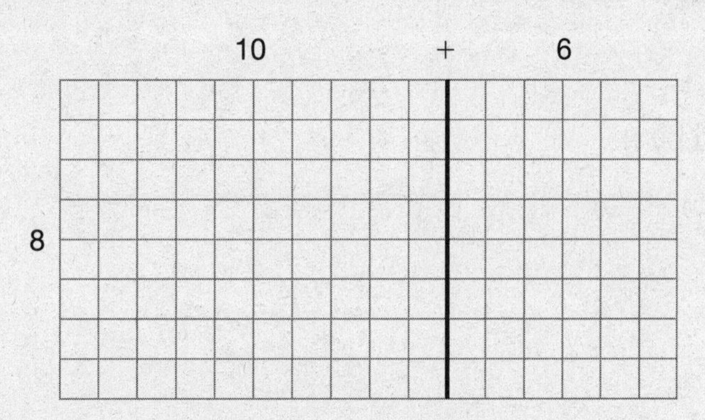

$8 \times 10 = 80$ and $8 \times 6 = 48$

Since $80 + 48 = 128$, $8 \times 16 = 128$

You can use the distributive property to help you solve multiplication word problems.

Example

There are 4 fourth-grade classes in Cherry Lane Elementary School. There are 26 students in each fourth-grade class. How many fourth-grade students are in Cherry Lane Elementary School?

The problem is asking for the total number of fourth-grade students in Cherry Lane Elementary School.

Since there are 4 equal-size groups of fourth-grade students, you should multiply to find the total number of fourth-grade students.

Write a multiplication equation that represents the information in the problem.

$4 \times 26 =$ total number of fourth-graders in the school

$4 \times 26 = \square$

Use the distributive property to solve the problem.
Rename the 2-digit factor as a sum.

$4 \times (20 + 6)$

Multiply each addend by 4.

$(4 \times 20) + (4 \times 6)$

Add the products.

$80 + 24 = 104$

So $4 \times 26 = 104$.

There are 104 fourth-grade students in Cherry Lane Elementary School.

 Practice

Directions: For questions 1 through 5, use the distributive property to fill in the blanks.

1. $29 \times (8 + 6) = (29 \times 8) + (29 \times \_\_\_\_\_\_)$

2. $17 \times (\_\_\_\_\_\_ + 4) = (17 \times 5) + (17 \times 4)$

3. $\_\_\_\_\_\_ \times (2 + 7) = (11 \times 2) + (11 \times 7)$

4. $(82 \times \_\_\_\_\_\_) + (82 \times \_\_\_\_\_\_) = 82 \times (26 + 38)$

5. $45 \times (30 + 9) = (\_\_\_\_\_\_ \times 30) + (\_\_\_\_\_\_ \times 9)$

6. Show that $2 \times (5 + 6) = (2 \times 5) + (2 \times 6)$.

7. Show that $(9 \times 3) + (9 \times 4) = 9 \times (3 + 4)$.

8. Which is the same as $2 \times (5 + 6)$?

   A. $(2 + 5) \times (2 + 6)$

   B. $(2 \times 5) + (2 \times 6)$

   C. $(2 \times 5) \times (2 \times 6)$

   D. $(2 + 5) + (2 + 6)$

9. Which is the same as $(8 \times 1) + (8 \times 3)$?

   A. $8 + (1 \times 3)$

   B. $8 \times (1 \times 3)$

   C. $8 \times (1 + 3)$

   D. $8 + (1 + 3)$

Directions: For questions 10 and 11, use the distributive property to solve each problem.

10. There are 42 boxes of books in a warehouse. There are 9 books in each box. How many books are there altogether?

____

11. The stars on a flag are in 18 rows and 8 columns. How many stars are there on the flag?

____

12. $7 \times 13 = ?$

Draw a model and use the distributive property to find the product.

Explain what the model shows and how you used the distributive property to find the product.

____

____

____

CCSS: 4.OA.3, 4.NBT.5

Lesson 10: Multiplying by a 1-Digit Number

Sometimes you want to multiply numbers greater than those in your multiplication table. You can follow the steps below to multiply a 2-digit number by a 1-digit number.

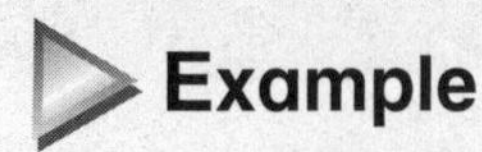

Example

Multiply: 9×35

Line up the digits that represent the same place values.

$$\begin{array}{r} 35 \\ \times\ 9 \\ \hline \end{array}$$

Multiply the 5 ones by 9: 9×5 ones $= 45$ ones.
Regroup 45 ones as 4 tens and 5 ones.
Write a 5 in the ones place of the product.
You can write a small 4 above the tens column in the problem, so you will remember to add the 4 tens to the other tens.

$$\begin{array}{r} {\scriptstyle 4}\ \\ 35 \\ \times\ 9 \\ \hline 5 \end{array}$$

Multiply the 3 tens by 9: 9×3 tens $= 27$ tens.
Add the 4 tens to the 27 tens: $4 + 27 = 31$.
Regroup the 31 tens as 3 hundreds and 1 ten.
Write a 1 in the tens place of the product and write a 3 in the hundreds place of the product.

$$\begin{array}{r} {\scriptstyle 4}\ \\ 35 \\ \times\ 9 \\ \hline 315 \end{array}$$

Therefore, $9 \times 35 = 315$.

The procedure is similar for multiplying a 3- or 4-digit number by a 1-digit number.

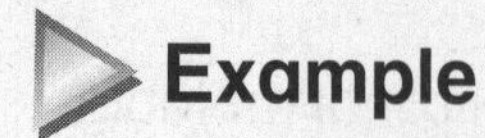

Example

Line up the digits with the same place values.

$$\begin{array}{r} 745 \\ \times\quad 6 \\ \hline \end{array}$$

Multiply the ones by 6: 6 × 5 ones = 30 ones.
Write a 0 in the ones place and regroup the 30 ones as 3 tens.

$$\begin{array}{r} {\scriptstyle 3} \\ 745 \\ \times\quad 6 \\ \hline 0 \end{array}$$

Multiply the tens by 6: 6 × 4 tens = 24 tens.
Then add the 3 tens to the 24 tens: 24 + 3 = 27.
Write a 7 in the tens place and regroup 20 tens as 2 hundreds.

$$\begin{array}{r} {\scriptstyle 23} \\ 745 \\ \times\quad 6 \\ \hline 70 \end{array}$$

Multiply the hundreds by 6: 6 × 7 hundreds = 42 hundreds.
Then add the 2 hundreds to the 42 hundreds: 42 + 2 = 44.
Write a 4 in the hundreds place. Write a 4 in the thousands place.

$$\begin{array}{r} {\scriptstyle 23} \\ 745 \\ \times\quad 6 \\ \hline 4470 \end{array}$$

So 6 × 745 = 4470.

Example

Line up the digits with the same place values.

$$\begin{array}{r} 2137 \\ \times \quad 3 \\ \hline \end{array}$$

Multiply the ones by 3: 3 × 7 ones = 21 ones.
Write a 1 in the ones place and regroup 20 ones as 2 tens.

$$\begin{array}{r} {}^{2} \\ 2137 \\ \times \quad 3 \\ \hline 1 \end{array}$$

Multiply the tens by 3: 3 × 3 tens = 9 tens.
Then add the 2 tens to the 9 tens: 9 + 2 = 11.
Write a 1 in the tens place and regroup 10 tens as 1 hundred.

$$\begin{array}{r} {}^{1\,2} \\ 2137 \\ \times \quad 3 \\ \hline 11 \end{array}$$

Multiply the hundreds by 3: 3 × 1 hundred = 3 hundreds.
Then add the 1 hundred to the 3 hundreds: 3 + 1 = 4.
Write a 4 in the hundreds place.

$$\begin{array}{r} {}^{1\,2} \\ 2137 \\ \times \quad 3 \\ \hline 411 \end{array}$$

Multiply the thousands by 3: 3 × 2 thousands = 6 thousands.
Write a 6 in the thousands place.

$$\begin{array}{r} {}^{1\,2} \\ 2137 \\ \times \quad 3 \\ \hline 6411 \end{array}$$

So 3 × 2137 = 6411.

Example

You can use place value and mental computation to multiply 2×3495.
Use expanded notation: $3495 = 3000 + 400 + 90 + 5$
Write an equation that shows multiplying each place value by 2.

$2 \times 3495 = (2 \times 3000) + (2 \times 400) + (2 \times 90) + (2 \times 5)$
So $2 \times 3495 = 6000 + 800 + 180 + 10$
and $2 \times 3495 = 6990$.

You can use multiplication to solve word problems involving greater numbers.

Example

A parking garage has 4 levels. Each level has 238 parking spaces. What is the total number of parking spaces in the parking garage?

The problem is asking for the total number of parking spaces in the parking garage.

Since there are 4 equal-sized levels of parking spaces, you should multiply to find the total number of parking spaces.

Write a multiplication equation that represents the information in the problem.

$4 \times 238 =$ total number of parking spaces in the parking garage

$4 \times 238 = \square$

To solve the problem, write the factors in vertical form. Line up the digits that represent the same place values.

$$\begin{array}{r} 238 \\ \times \quad 4 \\ \hline \end{array}$$

Multiply the ones, tens, and hundreds by 4.

$$\begin{array}{r} 238 \\ \times \quad 4 \\ \hline 952 \end{array}$$

So $4 \times 238 = 952$.

There are 952 parking spaces in the parking garage.

Practice

Directions: For questions 1 through 8, line up the digits that represent the same place values and then multiply.

1. $5 \times 37 =$ ______________

2. $3 \times 84 =$ ______________

3. $2 \times 276 =$ ______________

4. $7 \times 508 =$ ______________

5. $8 \times 359 =$ ______________

6. $6 \times 6219 =$ ______________

7. $4 \times 1853 =$ ______________

8. $9 \times 4022 =$ ______________

Directions: For questions 9 and 10, calculate each product mentally.

9. $22 \times 5 = ?$

   A. 55
   B. 100
   C. 105
   D. 110

10. $2 \times 348 = ?$

   A. 600
   B. 680
   C. 696
   D. 996

Solve each problem.

11. Talisha needs 476 tiles to tile the floors in her house. Each tile costs $7. What is the cost of the tiles? ____________

12. Danny rides his bike three days a week. He rides 32 kilometers on Mondays, 35 kilometers on Wednesdays, and 38 kilometers on Fridays. How many kilometers does Danny ride in 6 weeks? ____________

13. The principal of a local elementary school bought 8 desktop computers for the school's computer lab. Each computer cost $1189. How much did the principal pay for all of the computers? ____________

14. On average, a female hippopotamus weighs 3086 pounds and a male hippopotamus weighs 2204 pounds more than a female hippopotamus. On average, what is the total weight of 5 female and 5 male hippopotamuses?

____

Explain how you got your answer.

____

____

____

Lesson 11: Multiplying by a 2-Digit Number

To multiply by a 2-digit number, first multiply each digit in the other factor by the ones and by the tens. Then add the partial products.

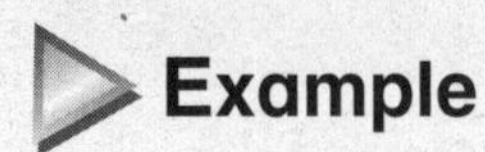

Example

Multiply: 18 × 24

Line up the digits that represent the same place values.
Then multiply the ones by 8: 8 × 4 ones = 32 ones.
Write a 2 in the ones place of the product and regroup 30 ones as 3 tens.

$$\begin{array}{r} {}^{3} \\ 24 \\ \times\ 18 \\ \hline 2 \end{array}$$

Multiply the 2 tens by 8: 8 × 2 tens = 16 tens.
Add tens: 16 tens + 3 tens = 19 tens.
Regroup as 1 ten and 9 ones. Write 9 in the tens place and 1 in the hundreds place.

$$\begin{array}{r} {}^{3} \\ 24 \\ \times\ 18 \\ \hline 192 \end{array}$$

← **8 × 24 = 192. 192 is the first partial product.**

Write a 0 under the 2 in 192 as a placeholder.
Multiply the ones by 1 ten: 1 ten × 4 ones = 4 tens.
Multiply the tens by 1 ten: 1 ten × 2 tens = 2 hundreds

$$\begin{array}{r} {}^{3} \\ 24 \\ \times\ 18 \\ \hline 192 \\ 240 \end{array}$$

← **10 × 24 = 240. 240 is the second partial product.**

Then add the partial products.

$$\begin{array}{r} {}^{3} \\ 24 \\ \times\ 18 \\ \hline 192 \\ +\ 240 \\ \hline 432 \end{array}$$

← **18 × 24 = 432, which is the sum of the partial products.**

Therefore, 8 × 24 = 432.

In the example on page 67, you multiplied the value of each digit in one factor first by the ones and then by the tens in the other factor. You can use the distributive property to see why this method of multiplying works.

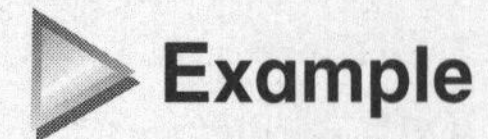

Example

$$\begin{array}{rl} 24 & \\ \underline{\times\ 18} & \\ 192 & 8 \times 24 \\ \underline{+\ 240} & 10 \times 24 \\ 432 & 18 \times 24 \end{array}$$

$$\begin{aligned} 18 \times 24 &= (10 + 8) \times 24 \\ 18 \times 24 &= (10 \times 24) + (8 \times 24) \\ &= \quad 240 \quad + \quad 192 \\ 18 \times 24 &= 432 \end{aligned}$$

There are other ways to multiply that give the same answer.

Example

$$\begin{array}{rl} 24 & \\ \underline{\times\ 18} & \\ 32 & (8 \text{ ones} \times 4 \text{ ones} = 3 \text{ tens} + 2 \text{ ones}) \\ 160 & (8 \text{ ones} \times 2 \text{ tens} = 16 \text{ tens}) \\ 40 & (1 \text{ ten} \times 4 \text{ ones} = 4 \text{ tens}) \\ \underline{+\ 200} & (1 \text{ ten} \times 2 \text{ tens} = 2 \text{ hundreds}) \\ 432 & (2 \text{ hundreds} + 23 \text{ tens} + 2 \text{ ones}) \end{array}$$

There are different ways that you can make a multiplication problem easier to calculate mentally.

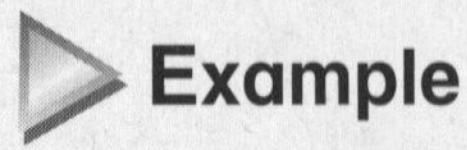

Example

Multiply: 53×10

Anytime you multiply a number by 10, simply write a "0" at the end of the number being multiplied by 10.

$53 \times 10 = 530$

Therefore, $53 \times 10 = 530$.

CCSS: 4.OA.3, 4.NBT.5

You can use multiplication to solve word problems involving greater numbers.

Example

Admission tickets at the Empire State Building cost \$20 for adults and \$14 for children. Last Saturday, 39 adults and 15 children visited the Empire State Building. How much did all of the admission tickets cost last Saturday?

The problem is asking for the total cost of admission tickets last Saturday. You can multiply to find the cost of all adult tickets and then multiply again to find the cost of all children's tickets. Then add to find the total cost.

Write equations that represent the information in the problem.

$39 \times \$20 = \text{cost of adults' tickets}$

$39 \times \$20 = \square$

Compute mentally. $39 \times \$2 = \$78$, so $39 \times \$20 = \$780$.

The cost of all adult tickets is \$780.

$15 \times \$14 = \text{cost of children's tickets}$

$15 \times \$14 = \square$

Write the factors in vertical form and multiply.

$$\begin{array}{r} \$14 \\ \times \ \ 15 \\ \hline \$210 \end{array}$$

$15 \times 14 = \$210$ The cost of all children's tickets is \$210.

$\$780 + \$210 = \text{total cost of all tickets}$

$\$780 + \$210 = \square$

Solve the problem.

$$\begin{array}{r} \$780 \\ +\ \$210 \\ \hline \$990 \end{array}$$

$\$780 + \$210 = \$990$

All of the admission tickets last Saturday cost \$990.

Practice

Directions: For questions 1 through 8, line up the digits with the same place values and then multiply.

1. 46 × 17 = ________

2. 51 × 24 = ________

3. 19 × 35 = ________

4. 62 × 47 = ________

5. 34 × 98 = ________

6. 27 × 15 = ________

7. 76 × 23 = ________

8. 57 × 44 = ________

Directions: For questions 9 and 10, compute each product mentally.

9. 64 × 10 = ?

   A. 460

   B. 600

   C. 630

   D. 640

10. 30 × 20 = ?

    A. 60

    B. 600

    C. 900

    D. 6000

CCSS: 4.OA.3, 4.NBT.5

11. In a movie theater, there are 42 rows of seats with 36 seats in each row. How many seats are there in the theater?

1512 seats

12. Admission tickets to the Science Museum cost $16 for school children and $21 for adults. A fourth-grade class of 12 girls and 14 boys, their teacher, and two parents go to the Science Museum on a field trip. What is the total cost of admission?

$459

13. A calculator costs $19. A laptop computer costs 20 times as much as the calculator. How much does the laptop computer cost?

1350

14. Mario has a stamp album. In the album, there are 28 pages with 15 stamps on each page, 14 pages with 18 stamps on each page, and 34 pages with 20 stamps on each page. What is the total number of stamps in Mario's album?

1352 stamps

28 × 15 = 420; 18 × 14 = 252; 34 × 20 = 680

680 + 420 + 252 = 1352

Explain how you got your answer.

I multiplyed 28x15, 14x18, 34x20 and added the products to get the sum

Lesson 12: Division Facts

The answer to a division problem is the **quotient**. The number you are dividing by is the **divisor**. The number being divided is the **dividend**. The number that is left over, if there is one, is the **remainder**. Here are two ways to write division facts:

quotient → 4
divisor → $6\overline{)24}$ ← **dividend**

24 ÷ 6 = 4 ← **quotient**
(24 is the **dividend**, 6 is the **divisor**)

Division Strategies

Grouping can be used to divide. Divide the total number into groups of the same size. If there is anything left over, it is the remainder. You can also use **repeated subtraction** to show the groupings.

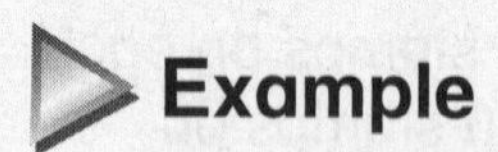

Example

Divide 13 nickels into groups of 4.

There are 3 groups of 4, and there is 1 left over.

13 ÷ 4 = 3 R1

Use repeated subtraction to show the number of groups.

13 − 4 = 9 **(1 time)**
9 − 4 = 5 **(2 times)**
5 − 4 = 1 **(3 times)**

There are 3 groups of 4, and there is 1 left over.

13 ÷ 4 = 3 R1

You can also use **equal sharing** to divide. Place one object in each group. Repeat this until all the objects are in a group. The number of objects in each group is the quotient.

Example

Claire has 9 pizza slices left over from a party. She wants to share them with 2 friends and have some for herself, so she needs to divide the pizza slices into 3 equal groups. How many pizza slices will be in each group?

Number the slices 1 through 3 until each slice has a number.

Count how many pizza slices have each number. This is how many slices will be in each group, or the number of slices that each person will get.
There are three 1s, three 2s, and three 3s.
Each person will get 3 pizza slices. There will be 3 pizza slices in each group.

$9 \div 3 = 3$

You have already seen how to solve a multiplication problem by making equal-sized jumps on a number line. You can show the multiplication $4 \times 6 = 24$ by starting at 0 and making 4 jumps of 6 tick marks each. The number line shows that 4 jumps forward of 6 tick marks in each jump is the same as starting at 0 and adding 4 groups of 6.

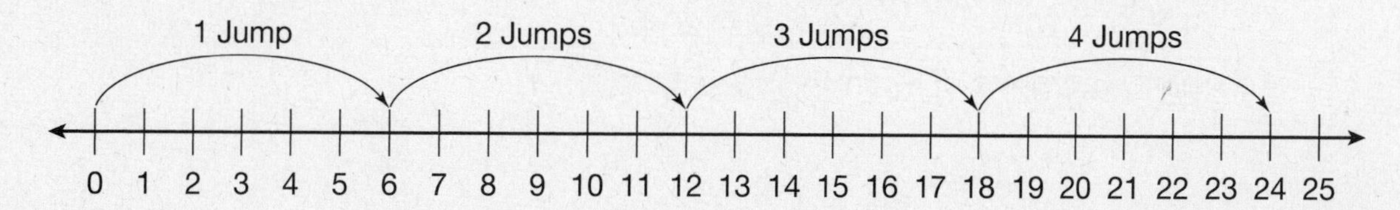

You can also solve a division problem by making equal-sized jumps on a number line. The total number being divided is the starting point on the number line. The number of tick marks in each jump is the known factor. The number of equal-sized jumps it takes to reach 0 is the unknown factor (the quotient).

Example

When you divide 24 pears into 6 groups having the same number of pears, how many pears will there be in each group?

Use equal jumps on a number line to solve the division problem.

Solve: $24 \div 6 = ?$

Start at 24. Make backward jumps of 6 tick marks each.
Count how many jumps it takes to get back to 0.

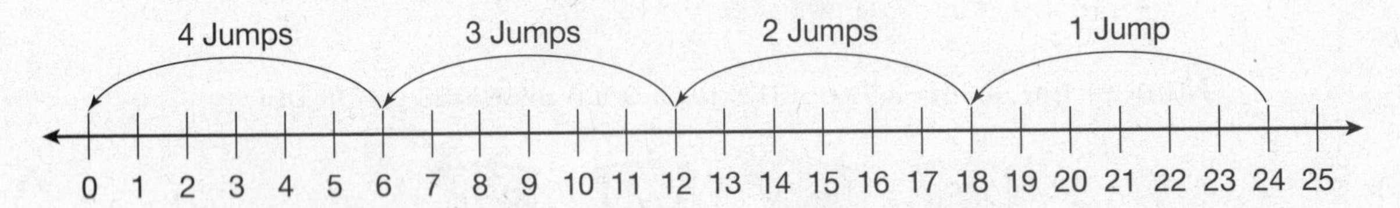

The number line shows that 4 jumps backward of 6 tick marks in each jump is the same as subtracting 4 groups of 6 from 24. When 24 pears are divided into 6 equal-sized groups, each group will have 4 pears, so $24 \div 6 = 4$.

The multiplication $4 \times 6 = 24$ and the division $24 \div 6 = 4$ are opposites because they "undo" one another. Every multiplication in which the factors are greater than 0 has a related division that is its opposite. These opposites are called **inverse operations**. When you apply the commutative property of multiplication to $4 \times 6 = 24$, the result is $6 \times 4 = 24$. The opposite division of $6 \times 4 = 24$ is $24 \div 4 = 6$. These four operations, which involve the numbers 4, 6, and 24, form a **multiplication and division fact family**.

Example

What is the multiplication and division fact family that can be formed from 4, 5, and 20?

$4 \times 5 = 20$	$5 \times 4 = 20$
$20 \div 5 = 4$	$20 \div 4 = 5$

CCSS: 4.OA.2, 4.NBT.6

You can use fact families to find an unknown factor or quotient in multiplication and division problems.

Example

What is the unknown factor?

$42 \div \square = 7$

Find out what other number is in the fact family by dividing: $42 \div 7 = 6$.

The three numbers in the fact family are 6, 7, and 42.
So $\square = 6$ and $42 \div \mathbf{6} = 7$.

Example

What is the unknown factor?

$8 \times \square = 72$

Find out what other number is in the fact family by dividing: $72 \div 8 = 9$.

The three numbers in the fact family are 8, 9, and 72.
So $\square = 9$ and $8 \times \mathbf{9} = 72$.

Practice

1. Divide the toothbrushes into groups of 4. How many groups of 4 are there? Are there any left over?

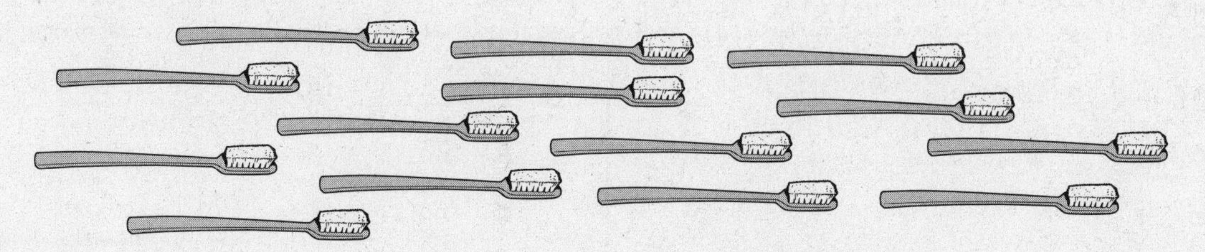

There are ______________ groups of 4, and there are ______________ left over.

2. In the space below, use repeated subtraction to find the answer to $25 \div 5$.

3. Divide the books into 5 equal groups. How many books are in each group?

There are ______________ books in each group.

4. Use the number lines to show the multiplication sentence $7 \times 3 = 21$ and its opposite sentence. Below each number line, write the number sentence that it represents.

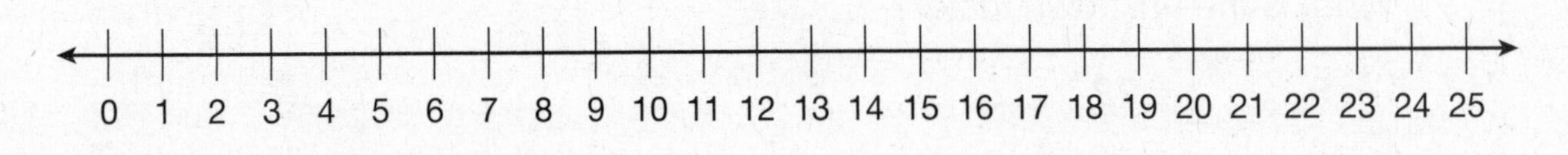

____

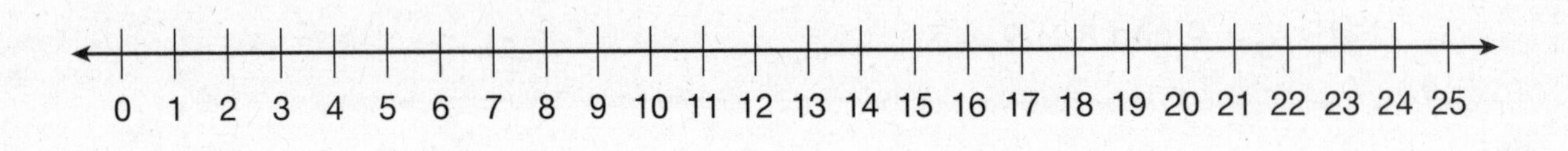

____

Directions: Use fact families to help you find the missing numbers for questions 5 and 6.

5. $36 \div$ ____ $= 9$
   - A. 3
   - B. 4
   - C. 5
   - D. 6

6. $7 \times$ ____ $= 49$
   - A. 5
   - B. 6
   - C. 7
   - D. 8

CCSS: 4.OA.2, 4.NBT.6

Directions: In Lesson 6, you made a multiplication table. Now you are going to use that table to solve division problems. On the left side of the table, find the number you are dividing by (the divisor). Read across that row until you see the number you are dividing (the dividend). Now look straight up that column to the number at the top of the table. That number is the answer (the quotient). Use the multiplication table you completed on page 45 to find the answers to questions 7 through 14.

7. $96 \div 12 =$ ____________

8. $60 \div 5 =$ ____________

9. $9 \div 3 =$ ____________

10. $27 \div 3 =$ ____________

11. $48 \div 8 =$ ____________

12. $70 \div 10 =$ ____________

13. $42 \div 6 =$ ____________

14. $121 \div 11 =$ ____________

Directions: For questions 15 and 16, fill in the blanks to complete the fact family.

15. $8 \times 5 = 40$ ____________

____________ ____________

16. ____________ ____________

____________ $63 \div 9 = 7$

17. Choose either question 15 or question 16. Tell which question you chose and explain how you found the answer.

____

____

____

Lesson 13: Dividing Greater Numbers

You can divide greater numbers by following a few steps.

Example

Divide: $57 \div 6$

Write the problem in long division form.

$6\overline{)57}$

Divide and multiply.

First, decide where to write the first digit of the quotient.

Think: How many 6s are in 57?

$9 \times 6 = 54$, and $54 < 57$.

$10 \times 6 = 60$, and $60 > 57$.

So 10 is too many 6s. Write a 9 in the ones place.

```
   9  ← How many 6s are in 57?
 6)57
 - 54  ← Multiply: 9 × 6 = 54
```

Subtract and write any remainder.

```
   9 R3
 6)57
 - 54
    3  ← The remainder (R) must be less than the divisor.
```

Therefore, $57 \div 6 = 9$ R3

Because multiplication and division are inverse operations, you can use multiplication to check the answer to a division problem. When you multiply the quotient by the divisor (and then add the remainder, if there is one), your answer should be the dividend.

```
   9
 × 6
  54
 + 3  ← Don't forget to add the remainder.
  57  ← This is the dividend, so your answer is correct.
```

Example

Mrs. Damon bakes 98 rolls. She divides them equally into 4 boxes for delivery to local restaurants. How many rolls does she put in each box? How many rolls are left over?
Write the problem in long division form.

$4\overline{)98}$

Divide and multiply.

First, decide where to write the first digit of the quotient.
Think: How many 4s are in 90?

$20 \times 4 = 80$, and $80 < 90$.

$30 \times 4 = 120$, and $120 > 90$.

So 30 is too many 4s. Write a 2 in the tens place.

$$\begin{array}{r} 2 \\ 4\overline{)98} \\ \underline{-\ 8} \end{array}$$

← **How many 4s are in 90?**

← **Multiply: 2 × 4 = 8**

Subtract and bring down the next number.

$$\begin{array}{r} 2 \\ 4\overline{)98} \\ \underline{-\ 8\downarrow} \\ 18 \end{array}$$

Divide, multiply, and subtract again.

$$\begin{array}{r} 24\ \text{R2} \\ 4\overline{)98} \\ \underline{-\ 8\downarrow} \\ 18 \\ \underline{-\ 16} \\ 2 \end{array}$$

← **Multiply: 4 × 4 = 16**

← **The remainder (R) must be less than the divisor.**

Use multiplication to check your answer. Multiply the quotient and the divisor. Then add the remainder.

$$\begin{array}{r} 24 \\ \underline{\times\ \ 4} \\ 96 \\ \underline{+\ \ 2} \\ 98 \end{array}$$

← **Don't forget to add the remainder.**

← **This is the dividend, so your answer is correct.**

Mrs. Damon puts 24 rolls in each box. There are 2 rolls left over.

Example

Divide: 225 ÷ 7

Write the problem in long division form.

$$7\overline{)225}$$

Divide and multiply.

First, decide where to write the first digit of the quotient.

Think: How many 7s are in 220?

30 × 7 = 210, and 210 < 220.

40 × 7 = 280, and 280 > 220.

So 40 is too many 7s. Write a 3 in the tens place.

```
    3     ← How many 7s are in 220?
 7)225
  - 21    ← Multiply: 3 × 7 = 21
  ----
```

Subtract and bring down the next number.

```
    3     ← The 3 means 3 tens, so it is in the tens place of the quotient.
 7)225
  - 21↓
  -----
    15
```

Divide, multiply, and subtract again.

```
   32
 7)225
  - 21↓
  -----
    15
  - 14    ← Multiply: 2 × 7 = 14
  -----
     1    ← The remainder (R) must be less than the divisor.
```

Therefore, 225 ÷ 7 = 32 R1.

Use multiplication to check your answer. Multiply the quotient and the divisor. Then add the remainder.

```
   32
 ×  7
 ----
  224
 +  1    ← Don't forget to add the remainder.
 ----
  225    ← This is the dividend, so your answer is correct.
```

CCSS: 4.OA.3, 4.NBT.6

Example

Mr. Chan has 1035 marbles to divide among his 6 children. How many marbles will each child get? How many marbles will be left over?
Write the problem in long division form.

$6\overline{)1035}$

Divide and multiply.

```
   1      ← How many 6s are in 1000?  Think: 100 × 6 = 600, and 600 < 1000.
6)1035                                 200 × 6 = 1200, and 1200 > 1000.
 − 6      ← Multiply: 1 × 6 = 6        So 200 is too many 6s.
```

Subtract and bring down the next number.

```
   1      ← The 1 stands for 1 hundred, so it is in the hundreds place of the quotient.
6)1035
 − 6↓
   43
```

Divide, multiply, subtract, and bring down the next number again.

```
   17
6)1035
 − 6↓
   43
 − 42↓    ← Multiply: 7 × 6 = 42
    15
```

Divide, multiply, and subtract again.

```
   172
6)1035
 − 6↓
   43
 − 42↓
    15
  − 12    ← Multiply: 2 × 6 = 12
     3    ← The remainder (R) must be less than the divisor.
```

You can use multiplication to check your answer. Multiply the quotient and the divisor. Then add the remainder.

```
  172
×   6
 1032
+   3   ← Don't forget to add the remainder.
 1035   ← This is the dividend, so your answer is correct.
```

Mr. Chan will give each of his children 172 marbles. There will be 3 marbles left over.

Sometimes you need to solve problems that have more than one step and involve two or more different operations.

Example

Molly went to a leather store where wallets were on sale for $23 each. She spent $124 for 2 wallets and 3 belts. If the belts she bought were all the same price, how much did she pay for each belt?

The problem is asking how much Molly paid for each belt.
Before you can find the price of a belt, you need to find how much of the total amount spent was spent on belts.

Write and solve an equation to represent how much Molly spent on belts.
You will multiply to find how much Molly spent on wallets.
You will subtract the amount spent on wallets from the total amount spent.

amount spent on belts = total spent − (number of wallets × cost per wallet)

$\square = 124 - (2 \times 23)$

$\square = 124 - 46$

$\square = 78$

So Molly spent $78 for 3 belts.

Now you can write and solve an equation to represent the amount Molly paid for each belt.

amount paid for 1 belt = amount spent on belts ÷ number of belts

$\square = 78 \div 3$

$\square = 26$

So each belt cost $26.

Practice

Directions: For questions 1 through 6, write the problem in long division form. Then find the quotient. Include the remainder if there is one.

1. 53 ÷ 6 = ______________

2. 75 ÷ 3 = ______________

3. 94 ÷ 5 = ______________

4. 19 ÷ 9 = ______________

5. 81 ÷ 7 = ______________

6. 27 ÷ 2 = ______________

7. If 78 chairs are arranged into equal rows of 6 chairs, how many rows will there be?

   A. 12

   B. 13

   C. 14

   D. 15

8. There are 4 small buses to take 33 boys and 31 girls on a field trip. If each bus carries the same number of students, how many students are on each bus?

   A. 10

   B. 12

   C. 14

   D. 16

Directions: For questions 9 through 16, write the problem in long division form. Then find the quotient. Include the remainder if there is one.

9. 217 ÷ 4 = ____________

10. 307 ÷ 5 = ____________

11. 837 ÷ 9 = ____________

12. 551 ÷ 8 = ____________

13. 297 ÷ 6 = ____________

14. 823 ÷ 7 = ____________

15. 130 ÷ 2 = ____________

16. 724 ÷ 3 = ____________

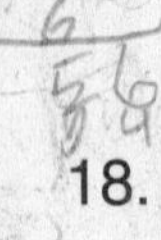

17. Mia has 116 tokens to use in playing games at an arcade. If each game takes 6 tokens, how many games can Mia play? How many tokens will she have left over?

____

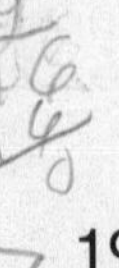

18. Next Sunday, 256 people will attend a wedding dinner. If each table can seat 8 people, how many tables will be needed?

____

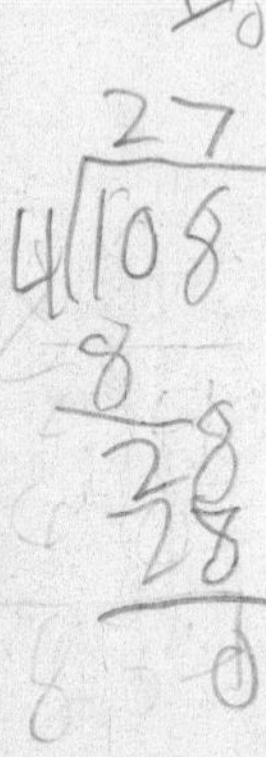

19. Ms. Myers's class of 29 students and Mr. Stimson's class of 32 students went on a joint field trip. The total cost of the trip was $962, which included the cost of $14 for each student. There were 4 adults accompanying the students. What was the cost for each adult?

____

Directions: For questions 20 through 25, write the problem in long division form. Then find the quotient. Include the remainder if there is one.

20. 5266 ÷ 7 = ________

21. 2183 ÷ 4 = ________

22. 3156 ÷ 6 = ________

23. 7190 ÷ 8 = ________

24. 6793 ÷ 9 = ________

25. 2690 ÷ 5 = ________

26. A school ordered 1288 pencils. Each box holds 8 pencils. How many boxes of pencils will the school receive?

____

Use multiplication to check your answer.

Explain why multiplication is used to check a division problem.

____

____

Lesson 14: Interpreting Remainders

When solving a division problem, sometimes there is a difference between the dividend and the greatest multiple of the divisor that is less than or equal to the dividend. That difference, called the **remainder**, is always less than the divisor. If the dividend is a multiple of the divisor, there will be no remainder.

Examples

When 28 is divided by 4, the quotient is 7. There is no remainder because 28, the dividend, is the greatest multiple of 4, the divisor, that is less than or equal to the dividend; and $28 - 28 = 0$.

When 28 is divided by 5, the quotient is 5 and there is a remainder. The remainder is 3 because the greatest multiple of the divisor that is less than or equal to the dividend is 25; and $28 - 25 = 3$.

When solving a problem that has a remainder, it is important to know the meaning of the remainder. Sometimes you round the quotient to the next whole number, sometimes you drop the remainder and use the quotient, and sometimes the remainder is the answer.

Example

Miss Kohl has 14 paintbrushes for her art class to use for a group project. Each group needs 4 paintbrushes for the project. What is the greatest number of groups there can be in the class?

Divide the 14 paintbrushes into groups of 4. How many groups of 4 are there?

CCSS: 4.OA.3

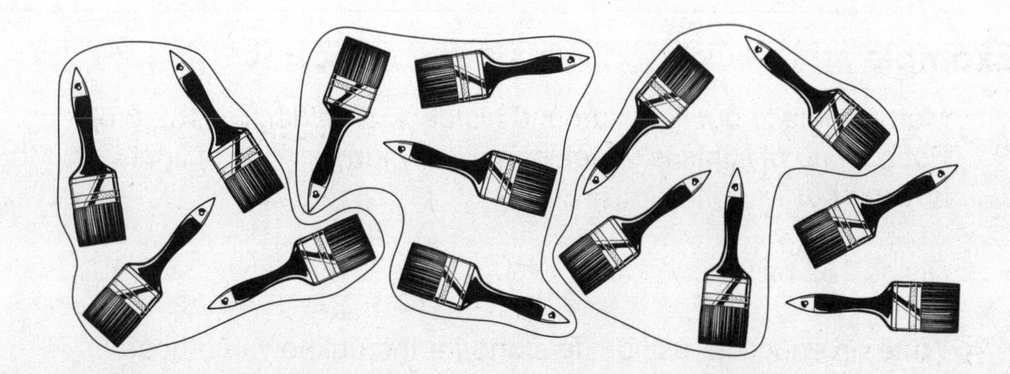

There are 3 groups of 4 paintbrushes, and there are 2 left over.
You can write an equation to represent the information in the problem. Use n to stand for the unknown number.

$14 \div 4 = n$

$14 \div 4 = 3 \text{ R2}$

In this problem, drop the remainder because 2 paintbrushes are not enough for a group to use.

There can be 3 groups in the class.

Example

Jenny is at an amusement park. She has $28. Each hot dog at the amusement park costs $5. She wants to buy the maximum number of hot dogs for the amount of money she has. How much money will Jenny have left over after buying the hot dogs?

Divide $28 by $5.

Write an equation, using n to stand for the unknown number.

$28 \div 5 = n$

$28 \div 5 = 5 \text{ R3}$

The answer shows that Jenny can buy 5 hot dogs, and she will have $3 left over. In this problem, the remainder is the answer.

Jenny will have $3 left over after buying the hot dogs.

Example

Marty wants to put lights around his deck. The deck is 152 feet long. Each string of lights is 9 feet long. How many strings of lights will Marty have to buy?

Divide 152 by 9.

Write an equation, using n to stand for the unknown number.

$152 \div 9 = n$

$152 \div 9 = 16 \text{ R}8$

In this problem, 16 strings of lights will not be enough, because Marty would still have 8 feet of deck without lights. Therefore, he must round up to the next whole number and buy another string of lights for the last 8 feet.
16 R8 rounded to the next whole number is 17.

Marty will have to buy 17 strings of lights.

Example

A clothing company has an order for 1067 pairs of jeans. The packing cartons hold 8 pairs of jeans. What is the minimum number of cartons needed to ship the entire order?

Divide 1067 by 8.

Write an equation, using n to stand for the unknown number.

$1067 \div 8 = n$

$1067 \div 8 = 133 \text{ R}3$

Decide which of the three types of remainder situation applies. In this problem, the question asks you to find the minimum number of cartons needed to ship all of the jeans. The remainder represents 3 pairs of jeans not packed, so 133 cartons will not be enough. Therefore, you need to round up to the next whole number and use another carton for the remaining 3 pairs of jeans.
133 R3 rounded to the next whole number is 134.

So 134 cartons are needed to ship the entire order.

TIP: *Left over*, *extra*, and *remaining* are clue phrases that the problem is asking for the remainder. *Maximum* and *at most* are clue phrases for dropping the remainder. *Minimum* and *at least* are clue phrases for rounding up to the next whole number.

Practice

Directions: For questions 1 through 10, find each answer. Then write whether you rounded the quotient to the next whole number, you dropped the remainder, or the remainder is the answer.

1. Marcy needs 5 buttons for each shirt. How many buttons will be left over?

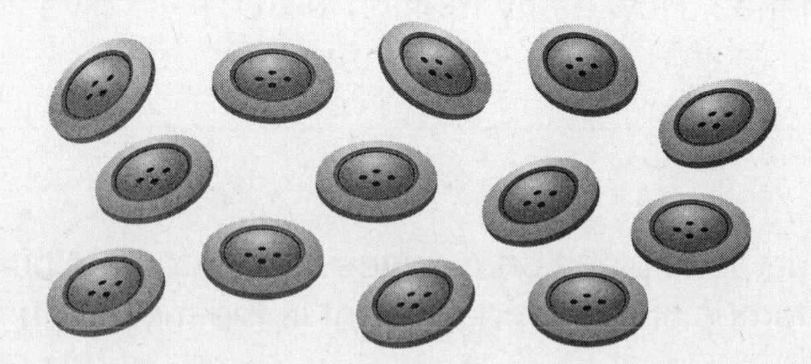

____

2. Steve has $142 to spend on lunches. He pays $5 each day for his lunch. How many days can Steve buy lunch?

____

3. Mrs. O'Brien bought 50 new books for the school library. If Mrs. O'Brien puts 9 books on each shelf, how many shelves will she need?

____

4. Jake has 40 pieces of candy. He is going to divide the candy equally into 3 bags to sell for charity. How many extra pieces of candy will Jake have?

____

5. Mindy wants to plant grass in her backyard. Her backyard is 100 square feet. The grass seed is sold in bags that hold enough grass seed to cover 8 square feet. How many bags of grass seed will Mindy have to buy in order to cover her backyard?

____

6. There are 1258 Army veterans signed up for a July 4 picnic. Each table will seat 9 veterans. How many tables will be needed?

____

7. LaToya baked 75 muffins. She kept 30 muffins for her family and divided the rest evenly among 7 friends. How many muffins will LaToya give to each friend?

____

8. Marlon has 146 stamps to place on 6 pages of his stamp album. If he places an equal number of stamps on each page, what is the maximum number of stamps on each page?

____

9. Benito made punch with 26 ounces of grape juice, 38 ounces of apple juice, and 52 ounces of lemon sparkling water. Then he made 8-ounce servings. How many ounces of punch were left over?

____

10. A florist put 31 flowers into 5 vases with the same number in each vase except one. How many extra flowers are in the vase with the greatest number of flowers?

____

Lesson 15: Estimating Products and Quotients

You can estimate products and quotients to help you make sure that your answer is **reasonable** (makes sense). If the answer is not close to the estimated answer, go back and look for an error in the computation. You can also estimate when an exact answer is not necessary.

There are several methods of estimating products and quotients, including rounding, compatible numbers, and clustering.

Rounding to Estimate Products

You can round numbers to the nearest 10 or 100 to estimate an answer to a multiplication problem.

Example

Find the exact product. Round each factor to the nearest 10 to estimate the product. Compare the estimate to the actual answer to see if the answer is reasonable.

Exact	Estimate
52	50
× 29	× 30
468	1500
+ 1040	
1508	

The estimated product is 1500. This is close to the actual product of 1508. Therefore, the answer is reasonable.

Example

Find the exact product. Double-check the answer. Keep the 1-digit factor and round the 4-digit factor to the nearest 10 to estimate the product. Then keep the 1-digit factor and round the 4-digit factor to the nearest 100 to estimate the product in a different way. Compare each estimate to the actual answer to see if the answer is reasonable.

Exact	Estimate	Estimate
4891	4890	4900
× 9	× 9	× 9
44,019	44,010	44,100

The estimated products are 44,010 and 44,100. These are close to the actual product of 44,019. Therefore, the answer is reasonable.

Example

In a library, there are 68 shelves with 22 books on each shelf. How many books are there in the library?

Step 1: **Find the actual product.**

$$\begin{array}{r} 68 \\ \times\ 22 \\ \hline 136 \\ +\ 1360 \\ \hline 1496 \end{array}$$

Step 2: **Estimate the product.**
Round 68 to 70. Round 22 to 20.

$70 \times 20 = 1400$

Step 3: **Compare the estimated product to the actual product.**
The estimated product is 1400. This is close to the actual product of 1496. Therefore, the exact answer is reasonable.

There are 1496 books in the library.

Using Compatible Numbers to Estimate Products

You can use **compatible numbers** to estimate an answer to a multiplication problem. Compatible numbers are numbers that make computations easier.

Examples

Estimate: 41×29

Change the 41 to 40 and the 29 to 30, since both are multiples of 10.

$40 \times 30 = 1200$

Using compatible numbers, the product is about 1200. (The actual product is 1189.)

Estimate: 52×31

Round 52 down to 50 and 31 down to 30. Use the fact $5 \times 3 = 15$ to multiply:

$50 \times 30 = 1500$. Since both factors were rounded down, the estimate of 1500 is less than the actual product. (The actual product is 1612.)

Estimate: 68×79

Round 68 up to 70 and 79 up to 80. Use the fact $7 \times 8 = 56$ to multiply:

$70 \times 80 = 5600$. Since both factors were rounded up, the estimate of 5600 is greater than the actual product. (The actual product is 5372.)

Clustering

You can use **clustering** when all of the numbers in a problem are close to the same number.

Example

Estimate: $9 \times 11 \times 9$

Notice that all of the numbers are close to 10. (All of the numbers **cluster** around 10.)

Change each number to 10 and multiply.

$10 \times 10 \times 10 = 1000$

Using clustering, the product is about 1000. (The actual product is 891.)

Rounding to Estimate Quotients

You can round numbers to the nearest 10 or 100 to estimate an answer to a division problem.

Example

Find the exact quotient. Round each number to the nearest 10 to estimate the quotient. Compare the estimate to the actual answer to see if the answer is reasonable.

Exact	Estimate
$\begin{array}{r} 9\text{ R6} \\ 9\overline{)87} \\ -81 \\ \hline 6 \end{array}$	$\begin{array}{r} 9 \\ 10\overline{)90} \end{array}$

The estimated quotient is 9. This is close to the actual quotient of 9 R6. Therefore, the answer is reasonable.

Example

Find the exact quotient. Round the 3-digit number to the nearest 100 and round the 1-digit number to the nearest 10 to estimate the quotient. Compare the estimate to the actual answer to see if the answer is reasonable.

Exact	Estimate
$\begin{array}{r} 32\text{ R1} \\ 8\overline{)257} \\ -24 \\ \hline 17 \\ -16 \\ \hline 1 \end{array}$	$\begin{array}{r} 30 \\ 10\overline{)300} \end{array}$

The estimated quotient is 30. This is close to the actual quotient of 32 R1. Therefore, the answer is reasonable.

Example

Find the exact quotient. Double-check the answer. Round both numbers to the nearest 10 to estimate the quotient. Then round the 4-digit number to the nearest 100 and round the 1-digit number to the nearest 10 to estimate the quotient in a different way. Compare each estimate to the actual answer to see if the answer is reasonable.

Exact: $9\overline{)3418}$ = 379 R7

```
  379 R7
9)3418
 -27
   71
  -63
   88
  -81
    7
```

Estimate: $10\overline{)3420}$ = 342

Estimate: $10\overline{)3400}$ = 340

The divisor in the estimate is greater than the actual divisor, so the actual quotient is greater than the estimated quotient. The estimated quotients are 342 and 340 and the actual quotient is 379 R7. Therefore, the estimate is close to the actual quotient and the answer is reasonable.

Example

Luis has a collection of 263 stamps. He puts 8 stamps on each page of his album. How many pages will have 8 stamps on them? How many stamps will be left over?

Step 1: **Find the actual quotient.**

```
  32 R7
8)263
 -24
   23
  -16
    7
```

Step 2: **Estimate the quotient.**
Round 263 to 300. Round 8 to 10.
$300 \div 10 = 30$

Step 3: **Compare the estimated quotient to the actual quotient.**
The estimated quotient is 30. This is close to the actual quotient of 32 R7. Therefore, the exact answer is reasonable.
Exactly 32 pages will have 8 stamps on them. There will be 7 stamps left over.

Using Compatible Numbers to Estimate Quotients

You can use **compatible numbers** to estimate an answer to a division problem. Compatible numbers are numbers that make computations easier.

Examples

Estimate: $58 \div 7$

Change the 58 to 56, since 56 is evenly divisible by 7.
$56 \div 7 = 8$

Using compatible numbers, the quotient is about 8.

Estimate: $238 \div 8$

Choose 240 to replace the dividend 238. Use a division fact to divide 240 by 8.
$24 \div 8 = 3$, so $240 \div 8 = 30$

The dividend used for the estimate is a little greater than the actual dividend.

Therefore, $238 \div 8$ is a little less than 30. (The actual quotient is 29 R6.)

Mental Division

You can make a division problem easier so that it can be calculated mentally.

Divide: $120 \div 5$
You can divide half of 120 by 5 and then double that answer.
Half of 120 is 60, so divide $60 \div 5$ mentally. $60 \div 5 = 12$
Now double your answer above. $2 \times 12 = 24$
Therefore, $120 \div 5 = 24$.

Practice

1. Which compatible numbers give the **best** estimate of the product of 19×48?

   A. 20×20

   B. 20×50

   C. 20×80

   D. 50×50

2. Which compatible numbers give the **best** estimate of the quotient of $29 \div 5$?

   A. $10 \div 5$

   B. $20 \div 5$

   C. $30 \div 2$

   D. $30 \div 5$

CCSS: 4.OA.3

Directions: For questions 3 through 6, find the exact answer. Then estimate the answer. Use the estimate to explain why the exact answer is reasonable.

3. 17 × 48 = ?

   actual 816 estimate ______

   ______

4. 4 × 2685 = ?

   actual 10740 estimate ______

   ______

5. 583 ÷ 7 = ?

   actual 83 R2 estimate ______

   ______

6. 1066 ÷ 4 = ?

   actual 266 R2 estimate ______

   ______

Directions: For questions 7 and 8, calculate each answer mentally and explain the strategy you used.

7. 97 × 10 = 970

   ______

   ______

8. 90 ÷ 5 = 18

   ______

   ______

Directions: For questions 9 through 12, find each answer. Then use estimation to see if your answer is reasonable. Tell which estimation method you used, and explain why you chose that method.

9. Benito earns $79 per day. How much does Benito earn in 28 days? ________

   estimate ____________

   estimation method

   ____

   ____

10. If 115 students are divided into equal teams of 5 students, how many teams will there be? ________

    estimate ____________

    estimation method

    ____

    ____

11. Mr. Anderson drives 1893 miles each month for his job as a traveling salesman. How many miles does Mr. Anderson drive in 6 months? ________

    estimate ____________

    estimation method

    ____

    ____

12. A large supermarket has 1418 juice boxes for sale. If the juice boxes are to be sold 8 to a pack, how many packs will there be? How many juice boxes are left over? ________

    estimate ____________

    estimation method

    ____

    ____

Lesson 16: Factors and Multiples

You have already learned that when you multiply a whole number by another whole number, the result is called a **multiple** of each of the two whole numbers. When you can divide a whole number by another whole number evenly (with no remainder), each of the two whole numbers is called a **factor** of the number it can divide. So, for example, 6 is a multiple of 3 and of 2; and 2 and 3 are factors of 6. Every whole number has an unlimited number of multiples.

Example

What are the multiples of 7?
You need to find the product of 7 and each of the whole numbers.

$7 \times 0 = 0$ $\quad$ $7 \times 3 = 21$

$7 \times 1 = 7$ $\quad$ $7 \times 4 = 28$

$7 \times 2 = 14$ $\quad$ $7 \times 5 = 35$ $\quad$ and so on . . .

Since 0 is a multiple of every number, you do not need to list 0 when you are listing the multiples of a number. The multiples of 7 are 7, 14, 21, 28, 35, and so on. You would never be able to list all of the multiples of 7.

You can skip count to list multiples of a number.

Example

Is 78 a multiple of 6?
Multiples of 6: 6, 12, 18, 24, 30, 36, 42, 48, 54, 60, 66, 72, 78, . . .
Yes. If you count until you either reach 78 or a number greater than 78, you will see that the list of multiples includes 78.

Example

Is 42 a multiple of 4?
Multiples of 4: 4, 8, 12, 16, 20, 24, 28, 32, 36, 40, 44, . . .
No, the list skips over 42 and goes from 40 to 44, so 42 is not a multiple of 4.

You can also divide to decide if one number is a multiple of another.

Example

Is 80 a multiple of 6?
Divide 80 by 6: $80 \div 6 = 13$ R2
Since there is a remainder, 6 does not divide 80 evenly.
So 80 is not a multiple of 6.

To find all the factors of a number, you need to find all the whole numbers that divide that number evenly. Factors always come in pairs. To find all the factor pairs of a number, find the two factors that, when multiplied, have that number as the product.

Example

What are all the factors of 6? What are the factor pairs of 6?
To find all the factors of 6, you need to find all the numbers that divide 6 evenly.

$6 \div \mathbf{1} = 6$	$6 \div \mathbf{3} = 2$	$6 \div 5 = 1$ R1
$6 \div \mathbf{2} = 3$	$6 \div 4 = 1$ R2	$6 \div \mathbf{6} = 1$

The numbers 1, 2, 3, and 6 are the factors of 6.

To find the factor pairs of 6, you need to find pairs of factors that have 6 as their product. There are two factor pairs of 6: one factor pair is 1 and 6; the other factor pair is 2 and 3.

Example

What are all the factors of 12? What are the factor pairs of 12?
You can use area models to show the factors and the factor pairs of 12.

1 (12) 2 (6)

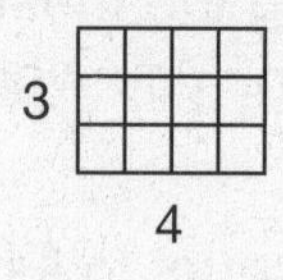

3 (4)

$1 \times 12 = 12$ $2 \times 6 = 12$ $3 \times 4 = 12$

The numbers 1, 2, 3, 4, 6, and 12 are the factors of 12. The factor pairs are 1×12, 2×6, and 3×4.

Example

What are all the factor pairs of 36?
Write multiplication sentences until the factors start to repeat. Ignore sentences that do not work.

$36 = 1 \times 36$
2×18
3×12
4×9
~~5 ×~~
6×6
~~7 ×~~
~~8 ×~~
9×4 ← **Stop. Repeat of 4 × 9.**

The factor pairs of 36 are 1×36, 2×18, 3×12, 4×9, and 6×6.

Example

What are all the factor pairs of 42?

Write multiplication sentences until the factors start to repeat. Ignore sentences that do not work.

$42 = 1 \times 42$

2×21

3×14

~~4 ×~~

~~5 ×~~

6×7

7×6 ← **Stop. Repeat of 6 × 7.**

The factor pairs of 42 are 1×42, 2×21, 3×14, and 6×7.

A **prime number** is a number greater than 1 that has exactly one factor pair.

Example

You can draw only one area model to show a prime number.

1

7

$1 \times 7 = 7$

The factor pair of 7 is 1×7.
7 is a prime number.

A **composite number** is a number greater than 1 that has two or more factor pairs.

Example

You can draw more than one area model to show a composite number.

1

8

$1 \times 8 = 8$

2

4

$2 \times 4 = 8$

The factor pairs of 8 are 1×8 and 2×4.
8 is a composite number.

Practice

Directions: For questions 1 through 4, list the first 10 multiples of the given number. Do not include 0 in the list.

1. 2

____

2. 3

____

3. 5

____

4. 9

____

5. Which number is a multiple of 8?

   A. 38
   B. 46
   C. 90
   D. 96

6. Which number is **not** a multiple of 4?

   A. 16
   B. 34
   C. 44
   D. 52

7. Which number is a multiple of both 4 and 6?

   A. 16
   B. 18
   C. 26
   D. 36

8. Which number is **not** a multiple of either 2 or 3?

   A. 12
   B. 21
   C. 25
   D. 30

CCSS: 4.OA.4

Directions: For questions 9 through 12, list the factors of the given number.

9. 58 ______________________________

10. 100 ______________________________

11. 24 ______________________________

12. 64 ______________________________

Directions: For questions 13 through 16, list the factor pairs of the given number.

13. 11 ______________________________

14. 44 ______________________________

15. 56 ______________________________

16. 71 ______________________________

17. Tito has a certain number of stickers. He wants to divide them into 2 or more equal groups. Which number of stickers **cannot** be divided into 2 or more equal groups?

    A. 31
    B. 33
    C. 35
    D. 39

18. Mandy has a certain number of marbles. She wants to divide them into 2 or more equal groups. Which number of marbles **can** be divided into 2 or more equal groups?

    A. 7
    B. 17
    C. 19
    D. 21

19. In the table below, the box for the number 1 has been shaded to remind you that 1 is neither prime nor composite. All the numbers that have 2 as a factor have been crossed out (other than 2 itself). Cross out all numbers (other than 3 itself) that have 3 as a factor, and then do the same for 4, 5, 6, 7, 8, and 9.

1	2	3	4	5	6	7	8	9	10
11	12	13	14	15	16	17	18	19	20
21	22	23	24	25	26	27	28	29	30
31	32	33	34	35	36	37	38	39	40
41	42	43	44	45	46	47	48	49	50
51	52	53	54	55	56	57	58	59	60
61	62	63	64	65	66	67	68	69	70
71	72	73	74	75	76	77	78	79	80
81	82	83	84	85	86	87	88	89	90
91	92	93	94	95	96	97	98	99	100

The numbers that are not crossed out are the prime numbers.

20. What are the prime numbers that are less than 100?

____

21. What does the table show that is special about the number 2?

____

Lesson 17: Dividing with Multiples of 10, 100, and 1000

You can use place value patterns and division facts to divide multiples of 10, 100, and 1000 by a 1-digit number.

Example

Find these quotients: $80 \div 4$ $800 \div 4$ $8000 \div 4$

Use the fact $8 \div 4 = 2$ and place value to find the quotients.

$80 \div 4 = ?$ Think: 8 tens $\div$ 4 = 2 tens, so $80 \div 4 = 20$.

$800 \div 4 = ?$ Think: 8 hundreds $\div$ 4 = 2 hundreds, so $800 \div 4 = 200$.

$8000 \div 4 = ?$ Think: 8 thousands $\div$ 4 = 2 thousands, so $8000 \div 4 = 2000$

Look for a pattern when multiples of 10, 100, or 1000 are divided by 10.

Think: tens divided by tens = ones
hundreds divided by tens = tens
thousands divided by tens = hundreds

Example

10 ÷ 10 = 1	**100** ÷ 10 = **10**	**1000** ÷ 10 = **100**
20 ÷ 10 = 2	**200** ÷ 10 = **20**	**2000** ÷ 10 = **200**
30 ÷ 10 = 3	**300** ÷ 10 = **30**	**3000** ÷ 10 = **300**

When dividing a multiple of 10, 100, or 1000 by 10, the quotient has 1 less zero than the dividend. The number of zeros in the quotient is determined by subtracting the number of zeros in the divisor from the number of zeros in the dividend. You can use this pattern, and basic division facts to help you find the quotient when you divide a multiple of 10, 100, or 1000 by a multiple of 10.

Example

Find these quotients: $80 \div 40$ $800 \div 40$ $8000 \div 40$

$80 \div 40 = ?$ $8 \div 4 = 2$ and **10** ÷ **10** = 1, so 8**0** ÷ 4**0** = 2

$800 \div 40 = ?$ $8 \div 4 = 2$ and **100** ÷ **10** = **10**, so 8**00** ÷ 4**0** = 2**0**

$8000 \div 40 = ?$ $8 \div 4 = 2$ and **1000** ÷ **10** = **100**, so 8**000** ÷ 4**0** = 2**00**

Be careful about writing zeros in the quotient when the fact you are using already has a zero in the dividend.

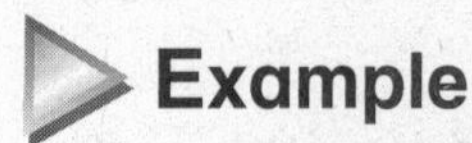

Example

Find the quotient: 3000 ÷ 60
Use place value and the division fact 30 ÷ 6 = 5.
Think: 30 hundreds divided by 6 tens = 5 tens, so 3000 ÷ 60 = 50.

Using the pattern and a basic division fact can help you find quotients without pencil and paper. If the dividend is a multiple of 10, you can use a shortcut to divide.

Example

360 ÷ 3 = ?

Step 1: Drop one or more zeros at the end of the dividend to make the number easier to divide.

36~~0~~ ÷ 3 = ?
36 ÷ 3 = ?

Step 2: Divide using the new dividend.

36 ÷ 3 = **12**

Step 3: Write the zero(s) you dropped in Step 1 at the end of the quotient.

12**0** ← Write one zero at the end of the quotient.

The answer is 120.

You have seen that the value of each place in a numeral is 10 times the value of the place next to it on the right. It is also true that the value of each place in a numeral divided by 10 is equal to the value of the place next to it on the right.

Example

Thousands	Hundreds	Tens	Ones
5	5	5	5
↑ The value of this 5 is 5 thousands, or 5000.	↑ The value of this 5 is 5 hundreds, or 500.	↑ The value of this 5 is 5 tens, or 50.	↑ The value of this 5 is 5 ones, or 5.

10 × 5 ones = 5 tens, so 10 × 5 = 50; and 50 ÷ 5 = 10.
10 × 5 tens = 5 hundreds, so 10 × 50 = 500; and 500 ÷ 50 = 10.
10 × 5 hundreds = 5 thousands, so 10 × 500 = 5000; and 5000 ÷ 500 = 10.

Practice

Directions: For questions 1 through 20, find the quotient.

1. 300 ÷ 5 = ______________
2. 600 ÷ 3 = ______________
3. 7200 ÷ 9 = ______________
4. 1200 ÷ 6 = ______________
5. 500 ÷ 5 = ______________
6. 420 ÷ 7 = ______________
7. 1600 ÷ 2 = ______________
8. 2000 ÷ 4 = ______________
9. 800 ÷ 8 = ______________
10. 2400 ÷ 3 = ______________
11. 70 ÷ 10 = ______________
12. 100 ÷ 10 = ______________
13. 8100 ÷ 90 = ______________
14. 8000 ÷ 40 = ______________
15. 540 ÷ 60 = ______________
16. 1000 ÷ 20 = ______________
17. 9000 ÷ 30 = ______________
18. 630 ÷ 70 = ______________
19. 640 ÷ 80 = ______________
20. 2000 ÷ 50 = ______________

Directions: For questions 21 and 22, use patterns and division facts to solve the problems.

21. 6400 ÷ 8 = ?

    A. 70
    B. 80
    C. 700
    D. 800

22. 4000 ÷ 10 = ?

    A. 4
    B. 40
    C. 400
    D. 4000

Directions: For questions 23 through 28, divide using the shortcut shown on page 106. Show your work in the space below each problem.

23. $240 \div 8 =$ ______________

24. $1800 \div 3 =$ ______________

25. $350 \div 7 =$ ______________

26. $8100 \div 9 =$ ______________

27. $2000 \div 4 =$ ______________

28. $450 \div 5 =$ ______________

29. $8000 \div 40 =$ ______________

    Explain how you got your answer.

    ____

    ____

30. Use the numeral 9999. Write a multiplication equation to describe how the value of the 9 in the hundreds place is related to the value of the 9 in the thousands place. Then write a division equation that represents this relationship.

    ____

    ____

    ____

CCSS: 4.OA.5

Lesson 18: Patterns

There are many kinds of patterns throughout the world. In this lesson, you will identify, describe, and extend geometric and number patterns. You will also use patterns to solve problems.

Geometric Patterns

Geometric patterns are made up of shapes. Look for a rule in the pattern. The rule describes how the pattern repeats. You can use the rule to help you find the next figure or the missing figure.

Sometimes the rule will involve the shapes of the figures.

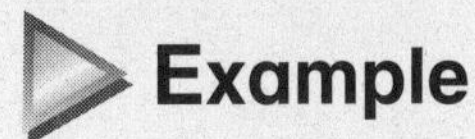

Example

Rule: triangle, square, star.

The next figure in the pattern is a **triangle**.

Sometimes the rule will involve the sizes of the figures.

Example

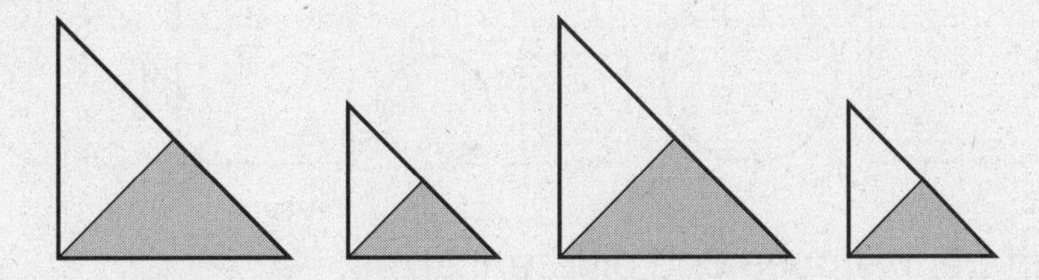

Rule: big triangle, small triangle.

The next figure in the pattern is a **big triangle**.

Sometimes the rule will involve the shadings or other markings.

Example

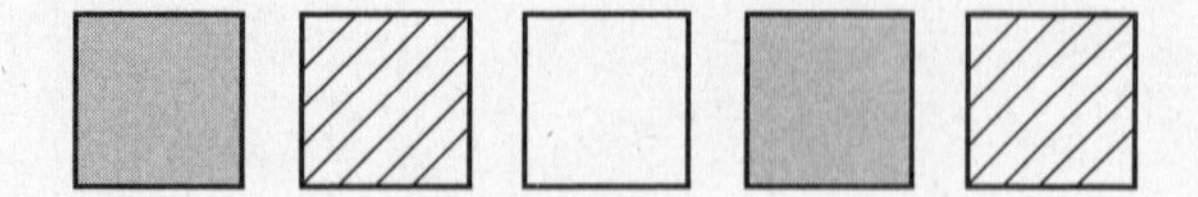

The figures are all same-size squares, but they are colored differently.

Rule: shaded square, striped square, blank square.

The next figure in the pattern is a **blank square**.

Sometimes the rule will involve the number or arrangement.

Example

Rule: Each figure has one more column of three stars than the figure before it.

The next figure in the pattern has **6 columns with 3 stars in each column**.

Sometimes the rule will be a combination of two or more features.

Example

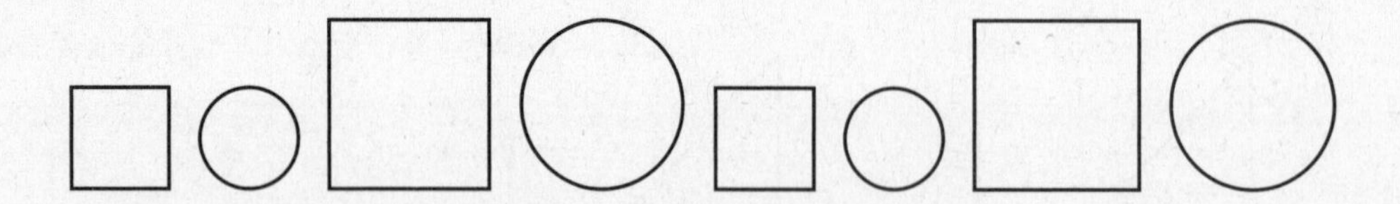

The figures are different shapes and different sizes.

Rule: small square, small circle, large square, large circle.

The next figure in the pattern is a **small square**.

Number Patterns

There are many kinds of number patterns. As with geometric patterns, you need to figure out the rule for a pattern to find a missing number or the next number. Any of the four mathematical operations (+, −, ×, ÷) can be used to make a pattern. A rule can be written using symbols or words as shown below.

Example

What is the rule for the pattern? What is the next number?

2, 8, 14, 20, 26, 32, 38, 44, . . .

In this pattern, the numbers increase by 6. The rule is **add 6**, or **+6**. Notice that the starting number is even, an even number is being added, and all the numbers in the pattern are even. The next number is 50.

Example

What is the rule for the pattern? What is the next number?

3, 7, 11, 15, 19, 23, 27, 31, . . .

In this pattern, the numbers increase by 4. The rule is **add 4**, or **+4**. Notice that the starting number is odd, an even number is being added, and all the numbers in the pattern are odd. The next number is 35.

Example

What is the rule for the pattern? What is the missing number?

3, 12, __?__, 192, 768, 3072

The numbers in this pattern are increasing rapidly. That means it's probably a multiplication pattern. The rule is **multiply by 4**, or **×4**. The missing number in the pattern is 48.

Sometimes a combination of operations is used as the rule for a pattern.

Example

What is the rule for the pattern? What is the next number?

1, 7, 4, 10, 7, 13, 10, 16, 13, . . .

In this pattern, the numbers increase by 6, then decrease by 3. The rule is **add 6, subtract 3**, or **+6**, **−3**. Notice that every other number increases by 3, and every other number alternates between odd and even. The next number is 19.

Sometimes you can find number patterns in tables.

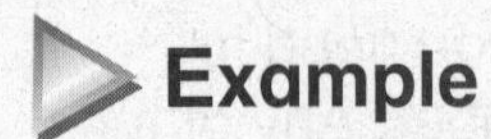

Example

What is the rule for the table, going from column x to column y?

x	y
1	6
3	8
5	10
7	12

Each number in column y is 5 more than the number in column x. The rule going from column x to column y is **add 5** or **+5**.

Number patterns in tables can be very helpful when you try to solve some real-world math problems.

Example

Rick saved the same amount of money each week for six weeks. The table below shows the total amount of money he has saved after each week. If Rick continues to save the same amount of money each week, how much money will he have saved after seven weeks?

Week	1	2	3	4	5	6	7
Total Amount Saved	\$3	\$6	\$9	\$12	\$15	\$18	?

Each number in the second row is 3 times the number in the first row. The rule going from the first row to the second row is **multiply by 3** or **×3**. So Rick will have saved 7×3, or \$21, after seven weeks.

CCSS: 4.OA.5

Practice

Directions: For questions 1 through 3, draw the next three figures for each pattern. Then, write the rule for the pattern.

1. 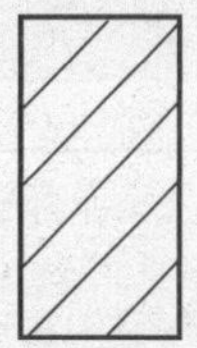   

What is the rule for the pattern? ______________________________

____

____

2. 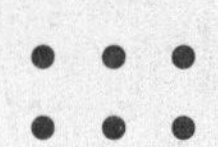  

What is the rule for the pattern? ______________________________

____

____

3.  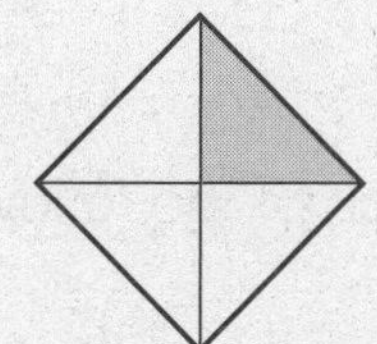 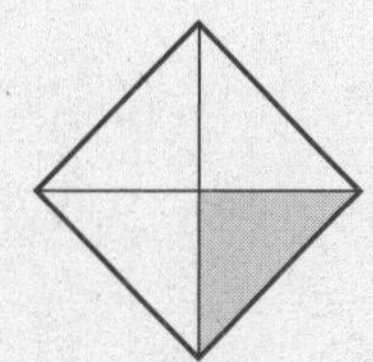

What is the rule for the pattern? ______________________________

____

____

Directions: For questions 4 through 8, fill in the missing numbers and write the rule for each pattern using symbols and numbers.

4. 5, 15, 45, _____, 405, _____, 3645 The rule is _______________.

5. 55, 53, 51, _____, 47, 45, 43, _____ The rule is _______________.

6. 4, 5, 50, _____, 510, 511, 5110, 5111 The rule is _______________.

7. 9216, 2304, _____, _____, 36, 9 The rule is _______________.

8. _____, 20, _____, 2000, 20,000 The rule is _______________.

9. Hannah is drawing the pattern below in the sand at the beach.

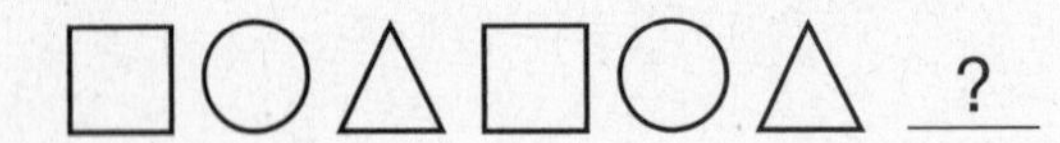

What figure comes next in this pattern?

A.

B.

C.

D.

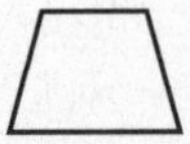

10. Renee wrote a pattern that follows the rule −9. Which of the following patterns could Renee have written?

A. 1, 2, 3, 4, 5

B. 1, 10, 19, 28, 37

C. 9, 8, 7, 6, 5

D. 47, 38, 29, 20, 11

CCSS: 4.OA.5

Directions: For questions 11 – 14, find the missing value in the table.

11.

x	*y*
1	7
4	28
7	49
11	

12.

x	*y*
3	15
4	16
7	19
9	

13.

x	*y*
99	49
87	37
74	
68	18

14.

x	*y*
2	18
5	
7	63
8	72

15. Miss Nelson's fourth graders put the following pattern on their bulletin board.

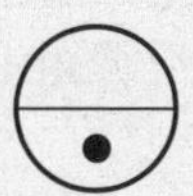 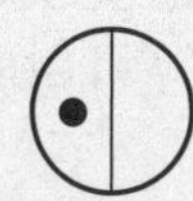 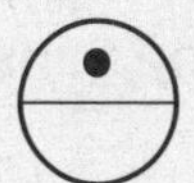 

Draw the next figure in the pattern.

Explain how you knew what to draw.

Directions: Use the information and table to answer questions 16 through 18.

Tasha planted the same number of flowers each week for seven weeks. She made the table to show the total number of flowers she had planted after each week.

Week	1	2	3	4	5	6	7
Total Number of Flowers	8	16	24				

16. What is the rule that shows how the number of flowers is related to the week number?

____

17. Fill in the missing numbers in the table.

18. If Tasha continues to plant the same number of flowers each week, how many flowers will she have after 10 weeks?

____

19. Clark was riding his bicycle down the sidewalk. He was looking at the address on each house as he went by. The first four addresses he saw were 2455, 2485, 2515, and 2545. What address do you think Clark will see next? Explain your answer.

____

____

20. What is the rule for the following pattern? What three numbers come next?
5, 8, 11, 14, 17, 20, 23, 26, 29, 32, . . .
Describe the numbers in the pattern in terms of their being odd or even.
What do you think is the reason for this?

____

____

____

Unit 2 Practice Test

Choose the correct answer.

1. $72 \div \underline{\ ?\ } = 8$

   A. 6

   B. 7

   C. 8

   D. 9

2. To find the product of 148×6, Lucinda did the easier multiplication 6×148. Which property of multiplication did Lucinda use?

   A. identity property

   B. associative property

   C. commutative property

   D. zero property

3. There are 5000 sheets of paper on a shelf next to a copying machine. The sheets of paper are arranged in equal stacks of 500 sheets each. How many stacks of paper are on the shelf?

   A. 5

   B. 10

   C. 50

   D. 100

4. $93 \times 37 = ?$

   A. 930

   B. 2441

   C. 3341

   D. 3441

5. A concert hall has 800 seats. For the concert last night, all the seats were filled. Each ticket to the concert cost $70. What was total amount of the ticket sales for the concert?

   A. $560

   B. $5600

   C. $56,000

   D. $560,000

6. $3657 \div 5 = ?$

   A. 731 R2

   B. 731 R3

   C. 731 R4

   D. 732

7. In solving $389 \div 7 = ?$, which equation will help you find the digit in the tens place of the quotient?

   A. $10 \times 7 < 380$

   B. $20 \times 7 < 380$

   C. $60 \times 7 > 380$

   D. $100 \times 7 > 380$

8. Latif has 27 DVDs. This is 3 times as many DVDs as Sarah has. In the following equations, the letter *s* stands for the number of DVDs Sarah has. Which equation matches the information given in the problem?

   A. $s = 3 \times 27$

   B. $3 \times s = 27$

   C. $3 \div s = 27$

   D. $s \div 3 = 27$

Use the information below to answer questions 9 and 10.

Cynthia has 2 quarters. Victor has 5 times as many quarters as Cynthia. Keiko has 5 more quarters than Cynthia has.

9. Use v for the number of quarters Victor has.

   Write and solve an equation for the number of quarters Victor has.

   ____

10. Use k for the number of quarters Keiko has.

    Write and solve an equation for the number of quarters Keiko has.

    ____

11. Solve the division problem $103 \div 5$. Complete the equation to show how you would check your answer.

    $103 = (5 \times \_\_\_\_\_) + \_\_\_\_\_$

12. How many squares are in the area model below?

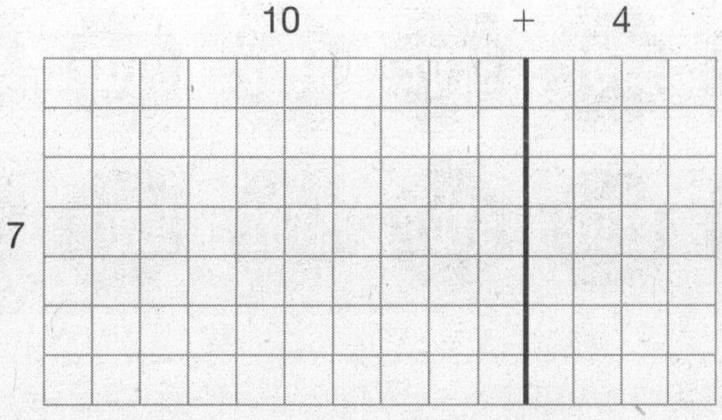

10 + 4

7

____

Explain how you found the answer.

____

____

____

13. What are the factor pairs of 52?

____

Explain how you found the answer.

____

____

____

14. Write two multiplication equations that are shown by the array below.

____

____

15. $33 \times 13 = 33 \times (10 +$ ______________ $)$

16. $510 \div 8 =$ ______________

17. What are the prime numbers between 20 and 50?

____

18. One fact in a fact family is $45 \div 9 = 5$. What are the other facts in the fact family?

____

19. $68 \times 19 =$ ______________

Explain how to use estimation to check your answer.

____

____

____

Solve each problem. Then use estimation to see if your answer is reasonable.

20. A student ticket to the science museum costs $11. How much would it cost for the 28 students in Xavier's class to buy tickets? __________

____

21. A grocer can fit 8 cans of soup into one carton. How many cartons are needed for 232 cans of soup? __________

____

22. Rolanda works for a moving company. Yesterday, she moved 19 stacks of 4 boxes and 21 stacks of 5 boxes. How many total boxes did Rolanda move yesterday? __________

____

23. A coat costs $81 and a sweater costs $9. How many times as much does the coat cost as the sweater? __________

____

24. There are 16 cups in 1 gallon. How many cups are in 51 gallons? __________

____

25. On Thursday, 98 students signed up for the town basketball league. On Friday, 73 more students signed up. Each team will have 9 players. Every student that signed up will be on a team. How many teams will there be in the league? __________

____

26. Karen needs 10 pounds of potting soil. She can only buy it in 3-pound bags. How many bags of soil does she need?

    Explain how you got your answer.

    ____

    ____

27. Draw the next figure in the pattern below.

    ______

28. Five winners of a contest get equal shares of the $1264 prize. How much money will be left over after the prize has been distributed?

    Explain how you got your answer.

    ____

    ____

29. Use the rule +5 to make a number pattern.

    Start with 1. Write the next 5 numbers.

    1, ____, ____, ____, ____, ____

    What do you notice about the numbers?

    ____

    ____

30. Heather has 149 trading cards. She keeps 24 of them and wants to split the rest equally among 5 friends. How many cards will each friend get?

Part A
What operations are needed to solve the problem?

____

Part B
How many cards will Heather have to give away after keeping 24 for herself?

____

Part C
What is your next step?

____

Part D
How many cards will each friend get?

____________________ cards

31. There are 4 fourth-grade classes in Cactus Road Elementary School. There are 28 students in each fourth-grade class. How many fourth-grade students are in Cactus Elementary School?

Part A
George used expanded notation to rename 28 as a sum. Then he used the distributive property and a place-value drawing to find the product. Show what George's drawing might look like.

Part B
Pia used expanded notation to rename 28 as a sum. Then she used the distributive property and an equation to find the product. Show the work Pia could have done.

Part C
How many fourth-grade students are in Cactus Road Elementary School?

____________________ students

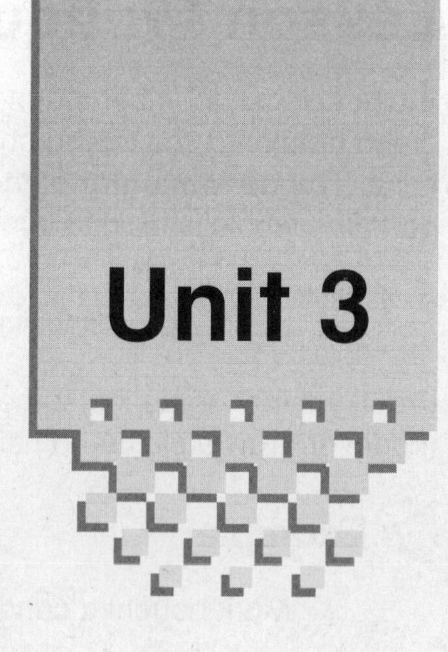

Number and Operations – Fractions

A fraction represents part of a whole or part of a group. Anytime you use a fraction, you are using part of a whole— $\frac{1}{2}$ of a sandwich, $\frac{3}{4}$ of an hour— or part of a group— $\frac{2}{5}$ of the marbles, $\frac{1}{3}$ of the students. As with whole numbers, you compute with fractions to solve problems that come up in everyday situations— problems that your knowledge of fractions helps you solve in the real world.

In this unit, you will find equivalent fractions, change mixed numbers to improper fractions, change improper fractions to mixed numbers, and compare fractions. You will also add and subtract fractions, and multiply fractions and whole numbers. You will study the connection between decimals and fractions. Finally, you will compare and order decimals.

In This Unit

Lesson 19: Equivalent Fractions

A **fraction** is a number that names parts of a whole or parts of a group. The **numerator** of the fraction is the top number. It tells you how many parts of the whole or group you have. The **denominator** of the fraction is the bottom number. It tells you how many parts the whole is divided into or how many parts are in the group.

numerator → $\frac{2}{5}$ ← **denominator**

Fractions that name the same amount are **equivalent fractions**. You can make a model or draw a picture to help you find equivalent fractions.

Example

Mark bought a candy bar and gave $\frac{5}{10}$ of it to his friend. Nadine bought the same kind of candy bar and gave $\frac{1}{2}$ of it to her friend. Who has more candy bar remaining?

$\frac{1}{2}$	$\frac{1}{2}$

$\frac{1}{10}$	$\frac{1}{10}$	$\frac{1}{10}$	$\frac{1}{10}$	$\frac{1}{10}$	$\frac{1}{10}$	$\frac{1}{10}$	$\frac{1}{10}$	$\frac{1}{10}$	$\frac{1}{10}$

They both have the same amount of candy bar remaining. The model shows that $\frac{5}{10}$ and $\frac{1}{2}$ are equivalent fractions.

Example

Mario and Abby each had the same size pizza. Mario's pizza was cut into 8 slices, and he ate 2 of those slices. Abby's pizza was cut into 4 slices, and she ate 1 of those slices. Who ate more pizza?

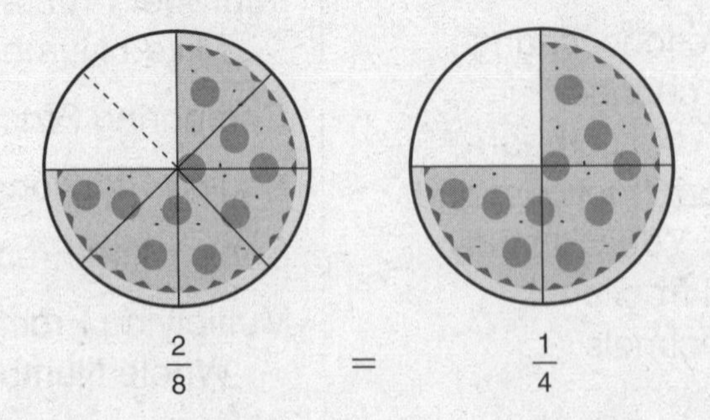

They both ate the same amount. The model shows that $\frac{2}{8}$ and $\frac{1}{4}$ are equivalent fractions.

CCSS: 4.NF.1

Example

Use the chart to find a fraction with a denominator of 4 that is equivalent to $\frac{1}{2}$. Explain how you know that the fractions are equivalent. Then look farther down the chart and find four more fractions that are equivalent to $\frac{1}{2}$.

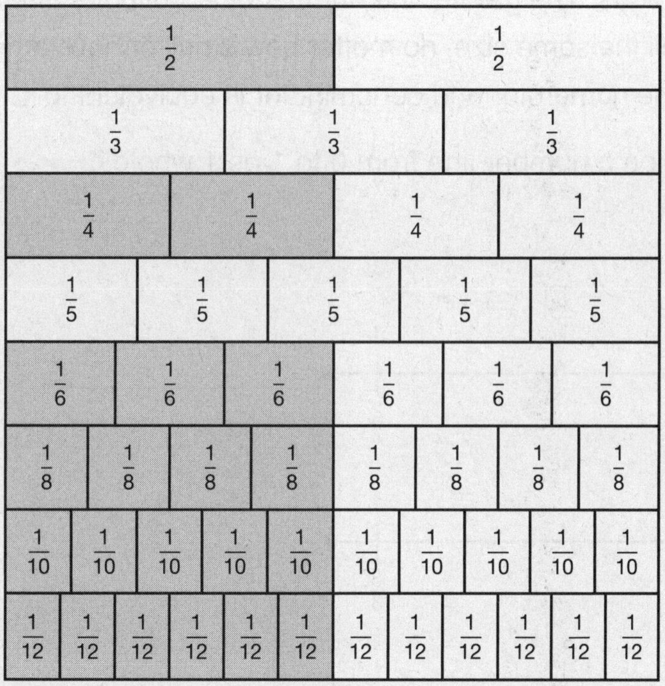

$\frac{1}{2}$ $\frac{1}{2}$

$\frac{1}{3}$ $\frac{1}{3}$ $\frac{1}{3}$

$\frac{1}{4}$ $\frac{1}{4}$ $\frac{1}{4}$ $\frac{1}{4}$

$\frac{1}{5}$ $\frac{1}{5}$ $\frac{1}{5}$ $\frac{1}{5}$ $\frac{1}{5}$

$\frac{1}{6}$ $\frac{1}{6}$ $\frac{1}{6}$ $\frac{1}{6}$ $\frac{1}{6}$ $\frac{1}{6}$

$\frac{1}{8}$ $\frac{1}{8}$ $\frac{1}{8}$ $\frac{1}{8}$ $\frac{1}{8}$ $\frac{1}{8}$ $\frac{1}{8}$ $\frac{1}{8}$

$\frac{1}{10}$ $\frac{1}{10}$ $\frac{1}{10}$ $\frac{1}{10}$ $\frac{1}{10}$ $\frac{1}{10}$ $\frac{1}{10}$ $\frac{1}{10}$ $\frac{1}{10}$ $\frac{1}{10}$

$\frac{1}{12}$ $\frac{1}{12}$ $\frac{1}{12}$ $\frac{1}{12}$ $\frac{1}{12}$ $\frac{1}{12}$ $\frac{1}{12}$ $\frac{1}{12}$ $\frac{1}{12}$ $\frac{1}{12}$ $\frac{1}{12}$ $\frac{1}{12}$

The model shows that the shaded section of the strip that shows 1 of 2 equal parts shaded is the same length as the shaded section of the strip that shows 2 of 4 equal parts shaded. So the fraction $\frac{1}{2}$ is equivalent to $\frac{2}{4}$.

The shaded section of the strip that shows 1 of 2 equal parts is also the same length as the shaded section of the strip that shows 3 of 6 equal parts shaded, 4 of 8 equal parts shaded, 5 of 10 equal parts shaded, and 6 of 12 equal parts shaded. So the fraction $\frac{1}{2}$ is also equivalent to $\frac{3}{6}$, $\frac{4}{8}$, $\frac{5}{10}$, and $\frac{6}{12}$.

You can divide 1 whole into any number of equal parts. When you divide 1 whole into 2 equal parts, each part is one half, or $\frac{1}{2}$, and the whole is two halves, or $\frac{2}{2}$. When you divide a whole into 3 equal parts, each part is one third, or $\frac{1}{3}$, and the whole is three thirds, or $\frac{3}{3}$. The pattern is the same for a whole divided into 4 equal parts, 5 equal parts, and so on. So the whole number 1 can have different names because the whole is divided into smaller or larger equal parts. The greater the number of equal parts, the smaller each part will be. The whole is still the same size, no matter how small or how large each part is. Any fraction with the same numerator and denominator is equivalent to 1.

You can think of the space on a number line from 0 to 1 as 1 whole.

Examples

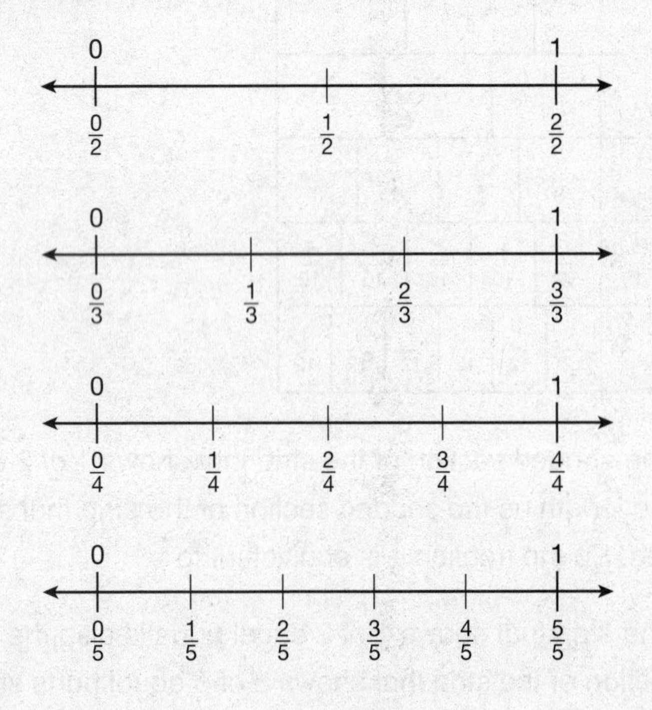

CCSS: 4.NF.1

You already know the identity property of multiplication. When you multiply any number by 1, the product is that number. This property is also true for fractions. When you multiply the numerator and denominator by the same number, you are multiplying the fraction by 1. You can use this property to change a fraction to an equivalent fraction.

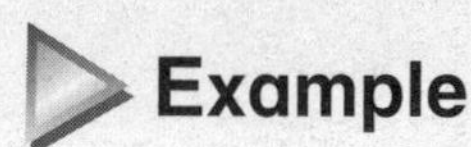

Example

How would you use the identity property of multiplication to change $\frac{1}{4}$ to an equivalent fraction that has a denominator of 8? How would you use it to change $\frac{1}{2}$ to an equivalent fraction that has a denominator of 6?

You know that 2 × **4** = 8. So multiply the denominator by 4 to get a denominator of 8. Then multiply the numerator by the same number.

$$\frac{1 \times ?}{2 \times 4} = \frac{?}{8}$$

$$\frac{1 \times 4}{2 \times 4} = \frac{4}{8}$$

You know that 2 × **3** = 6. So multiply the denominator by 3 to get a denominator of 6. Then multiply the numerator by the same number.

$$\frac{1 \times ?}{2 \times 3} = \frac{?}{6}$$

$$\frac{1 \times 3}{2 \times 3} = \frac{3}{6}$$

Therefore, just as you saw when you used the models of strips divided into different numbers of equal parts, the fractions $\frac{4}{8}$ and $\frac{3}{6}$ are each equivalent to $\frac{1}{2}$. When you multiply the numerator and denominator of a fraction by the same number, the result is a new fraction that is equivalent to the original fraction. This works because when you multiply by $\frac{4}{4}$ or $\frac{3}{3}$ or any other fraction that has the same numerator and denominator, you are really multiplying by 1. Even though you are changing the numerator and denominator, the value of the fraction stays the same.

Practice

Directions: Use the chart below to answer questions 1 through 4.

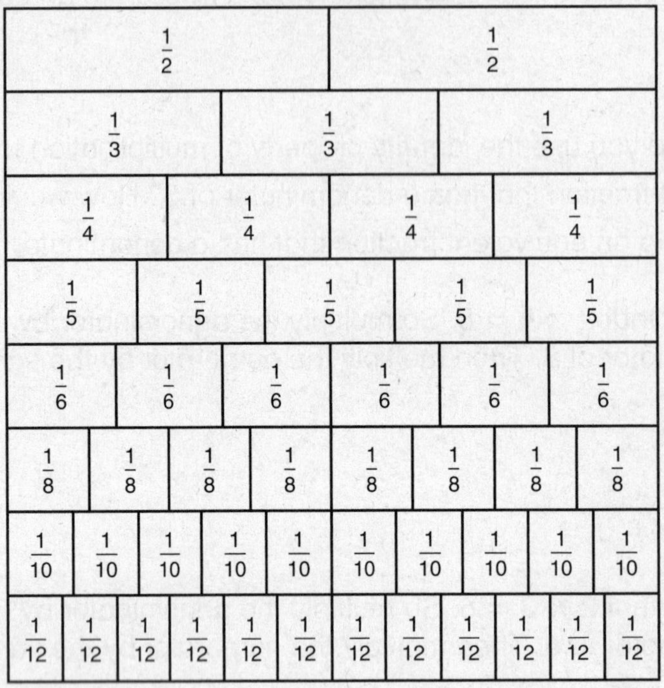

$\frac{1}{2}$	$\frac{1}{2}$										
$\frac{1}{3}$	$\frac{1}{3}$	$\frac{1}{3}$									
$\frac{1}{4}$	$\frac{1}{4}$	$\frac{1}{4}$	$\frac{1}{4}$								
$\frac{1}{5}$	$\frac{1}{5}$	$\frac{1}{5}$	$\frac{1}{5}$	$\frac{1}{5}$							
$\frac{1}{6}$	$\frac{1}{6}$	$\frac{1}{6}$	$\frac{1}{6}$	$\frac{1}{6}$	$\frac{1}{6}$						
$\frac{1}{8}$	$\frac{1}{8}$	$\frac{1}{8}$	$\frac{1}{8}$	$\frac{1}{8}$	$\frac{1}{8}$	$\frac{1}{8}$	$\frac{1}{8}$				
$\frac{1}{10}$	$\frac{1}{10}$	$\frac{1}{10}$	$\frac{1}{10}$	$\frac{1}{10}$	$\frac{1}{10}$	$\frac{1}{10}$	$\frac{1}{10}$	$\frac{1}{10}$	$\frac{1}{10}$		
$\frac{1}{12}$	$\frac{1}{12}$	$\frac{1}{12}$	$\frac{1}{12}$	$\frac{1}{12}$	$\frac{1}{12}$	$\frac{1}{12}$	$\frac{1}{12}$	$\frac{1}{12}$	$\frac{1}{12}$	$\frac{1}{12}$	$\frac{1}{12}$

1. Write the fractions that are equivalent to $\frac{1}{4}$.

   $\frac{1}{4} = \frac{2}{8} = \frac{3}{12}$

2. Write the fractions that are equivalent to $\frac{2}{3}$.

   $\frac{2}{3} = \frac{4}{6} = \frac{8}{12}$

3. Which fraction is equivalent to $\frac{1}{5}$?

   A. $\frac{2}{8}$

   B. $\frac{3}{8}$

   C. $\frac{2}{10}$

   D. $\frac{3}{12}$

4. Which fraction is **not** equivalent to $\frac{1}{2}$?

   A. $\frac{2}{4}$

   B. $\frac{3}{5}$

   C. $\frac{4}{8}$

   D. $\frac{6}{12}$

5. Shade in the squares to show a fraction equivalent to $\frac{5}{6}$.

6. Shade an equivalent fraction to the figure shown.

   Then complete the equation to show the equivalent fractions.

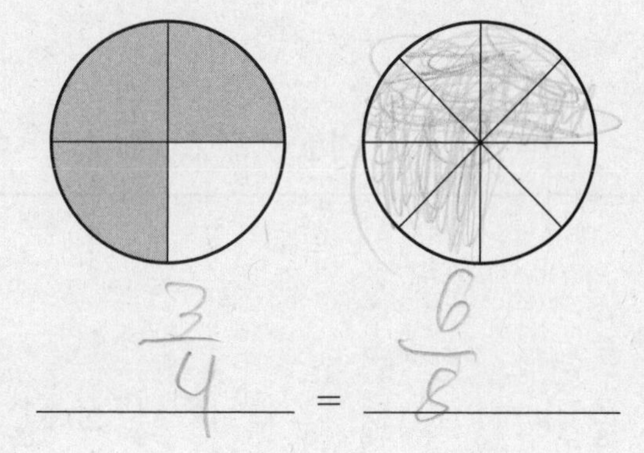

$\frac{3}{4}$ = $\frac{6}{8}$

7. Write an equation to show a fraction that is equivalent to $\frac{2}{5}$.

$\frac{2}{5} = \frac{2 \times ?}{5 \times ?} = \frac{?}{?}$ $\frac{2}{5} = \frac{2}{5} \times \frac{2}{2} = \frac{4}{10}$

Explain why the value of the fraction stays the same.

its equele bccaus I looked at the bar modle on Page 130

8. In the space below, draw a picture to show that $\frac{1}{3}$ is equivalent to $\frac{2}{6}$.

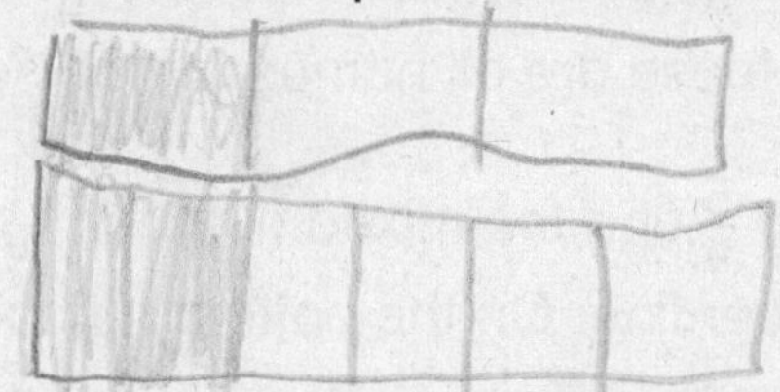

Explain how you know that the fractions are equivalent.

I Drew a bar for it and its equal

Lesson 20: Improper Fractions and Mixed Numbers

When numbers equal to or greater than 1 are represented as fractions, they are called **improper fractions**. An improper fraction is a fraction with a numerator that is greater than or equal to its denominator. Numbers greater than 1 that are not whole numbers can be represented either as improper fractions or as **mixed numbers**. A mixed number has a whole number part and a fraction part.

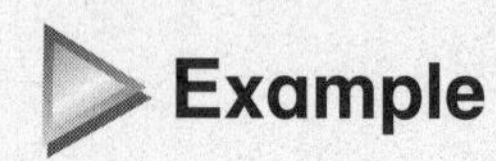

Example

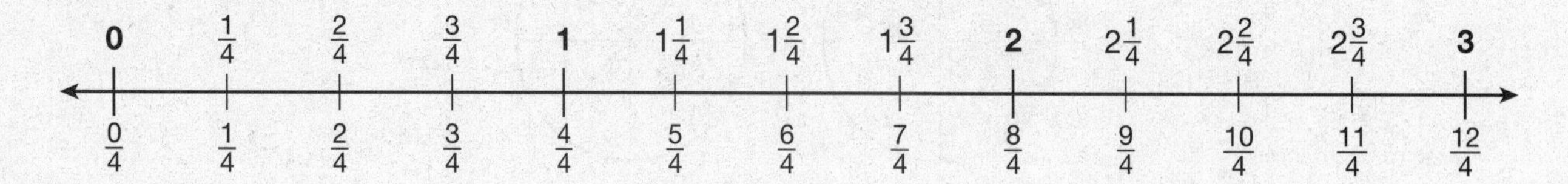

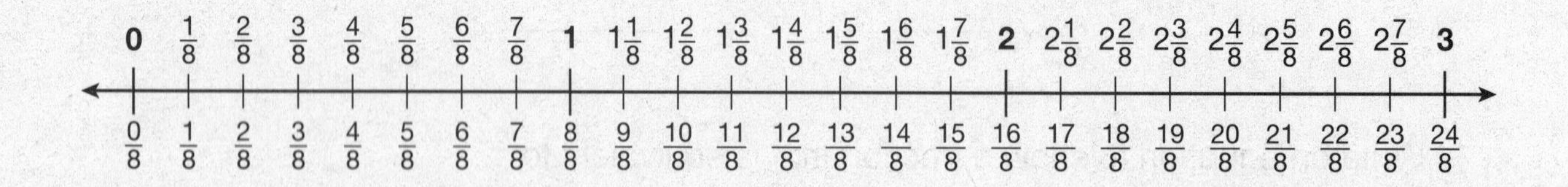

You will see that, on the number lines above, there is more than one name for any number that is equal to or greater than 1. The name above the point is either a whole number or a mixed number. The name below the point is an improper fraction. The point is still the same, whether it is represented by a mixed number, a whole number, or an improper fraction.

Example

What are three different names for the improper fraction $\frac{5}{4}$? How do the number lines above show that these are all names for the same number?

The top number line shows that $\frac{5}{4}$ and the mixed number $1\frac{1}{4}$ are both names for the same mark on the number line. On the bottom number line, the mixed number $1\frac{2}{8}$ is the same distance from 0 as $\frac{5}{4}$ is, so it is another name for the same number. Also on the bottom number line, $1\frac{2}{8}$ and the improper fraction $\frac{10}{8}$ are both names for the same mark on the number line.

Therefore, $1\frac{1}{4}$, $1\frac{2}{8}$, and $\frac{10}{8}$ are all different names for the improper fraction $\frac{5}{4}$.

Example

Use the number line to find an improper fraction that is equivalent to $2\frac{3}{5}$.

To find the improper fraction equivalent to $2\frac{3}{5}$, start at $\frac{0}{5}$ and count by fifths until you reach the point for the number of fifths that is labeled $2\frac{3}{5}$. The improper fraction that is equivalent to $2\frac{3}{5}$ is $\frac{13}{5}$.

Example

Use the number line to find a mixed number that is equivalent to $\frac{14}{5}$.

To find the mixed number equivalent to $\frac{14}{5}$, start at 0 and count by fifths until you reach $\frac{14}{5}$. Look at the numbers above the number line. You will see that $\frac{14}{5}$ is greater than 2 and less than 3. So start at 2 and count mixed numbers by fifths ($2\frac{1}{5}$, $2\frac{2}{5}$, and so on) until you reach the point for $\frac{14}{5}$. The mixed number that is equivalent to $\frac{14}{5}$ is $2\frac{4}{5}$.

Example

Use either of the number lines above to find an improper fraction that is equivalent to the whole number 2.

You already know that 1 is equivalent to $\frac{5}{5}$. So you can start at 1 and count by fifths until you reach 2. An improper fraction that is equivalent to 2 is $\frac{10}{5}$. Notice that whenever an improper fraction is equivalent to a whole number, its numerator can be divided evenly by its denominator.

To change a whole number to an equivalent improper fraction, multiply the whole number and the number you want to use as the denominator.

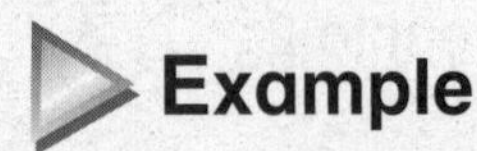

Example

Change the whole number 8 to an equivalent improper fraction. First, choose a denominator. Then multiply 8 by the denominator you choose. To change 8 to an improper fraction having a denominator of 10, multiply 8 × 10 to find the numerator.

$$8 = \frac{8 \times 10}{10} = \frac{80}{10}$$

You can use a shortcut to change a mixed number to an equivalent improper fraction.

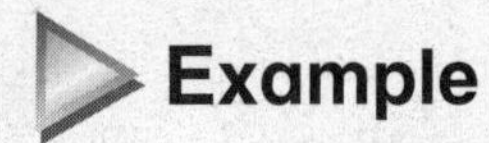

Example

Change $3\frac{3}{4}$ to an equivalent improper fraction.

To change a mixed number to an equivalent improper fraction, first multiply the whole number times the denominator of the fraction. Add the numerator of the fraction to the product. Write the result as the numerator of the improper fraction. Write the denominator of the fraction as denominator of the improper fraction.

$$3\frac{3}{4} = \frac{(3 \times 4) + 3}{4} = \frac{15}{4}$$

You can use a shortcut to change an improper fraction to an equivalent whole number or mixed number.

Example

Change $\frac{18}{4}$ to an equivalent mixed number.

To change an improper fraction to an equivalent whole number or mixed number, divide the numerator by the denominator. If there is a remainder, write it as the numerator of the fraction part of the mixed number. The denominator of the fraction will be the denominator of the improper fraction.

$$\frac{18}{4} = 18 \div 4 = 4\frac{2}{4}$$

Practice

Directions: For questions 1 through 8, write each improper fraction as a mixed number.

1. $\frac{14}{5}$ = ____________

2. $\frac{17}{2}$ = ____________

3. $\frac{31}{4}$ = ____________

4. $\frac{26}{6}$ = ____________

5. $\frac{11}{3}$ = ____________

6. $\frac{60}{8}$ = ____________

7. $\frac{5}{2}$ = ____________

8. $\frac{28}{3}$ = ____________

Directions: For questions 9 through 16, write each mixed number as an improper fraction.

9. $3\frac{1}{2}$ = ____________

10. $2\frac{5}{8}$ = ____________

11. $1\frac{9}{10}$ = ____________

12. $2\frac{2}{5}$ = ____________

13. $5\frac{3}{4}$ = ____________

14. $4\frac{3}{8}$ = ____________

15. $6\frac{5}{6}$ = ____________

16. $5\frac{1}{3}$ = ____________

17. Which is a way to write 3 as an improper fraction?

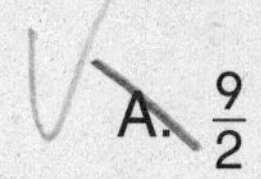

A. $\frac{9}{2}$

B. $\frac{6}{3}$

C. $\frac{8}{4}$

D. $\frac{12}{4}$

18. What is $\frac{25}{2}$ written as a mixed number?

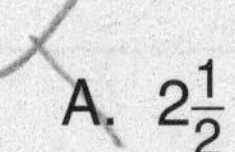

A. $2\frac{1}{2}$

B. 12

C. $12\frac{1}{2}$

D. $25\frac{1}{2}$

Directions: For questions 19 through 22, name the two whole numbers each improper fraction is between. Then name the whole number the improper fraction is closer to.

19. $\frac{15}{8}$ is between __________ and __________. It is closer to __________.

20. $\frac{19}{6}$ is between __________ and __________. It is closer to __________.

21. $\frac{27}{4}$ is between __________ and __________. It is closer to __________.

22. $\frac{36}{5}$ is between __________ and __________. It is closer to __________.

23. Use the number line to find an improper fraction that is equivalent to $1\frac{5}{6}$.

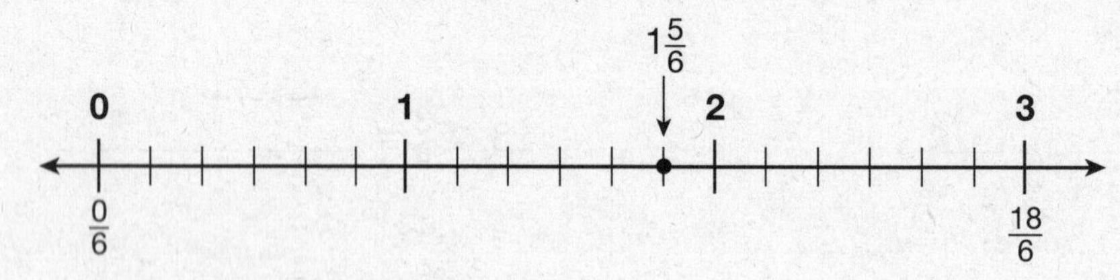

__________ is equivalent to $1\frac{5}{6}$.

Explain how you used the number line to get your answer.

____

____

24. Use the number line to find a mixed number that is equivalent to $\frac{7}{3}$.

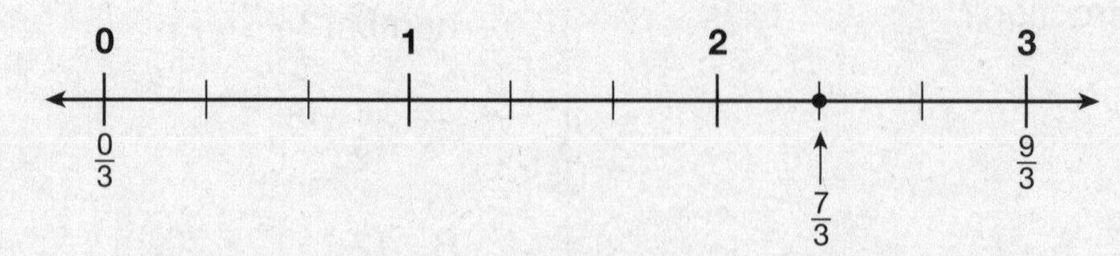

__________ is equivalent to $\frac{7}{3}$.

Explain how you used the number line to get your answer.

____

____

CCSS: 4.NF.2

Lesson 21: Comparing Fractions

You already know the following about two fractions that are parts of the same-size wholes: If the fractions have different numerators but the same denominator, the greater fraction is the one with the greater numerator. If the fractions have different denominators but the same numerator, the greater fraction is the one with the smaller denominator.
In this lesson, you will compare fractions that have both different numerators and different denominators. Remember: For a comparison of two fractions to be meaningful, the fractions must be parts of the same-size wholes.

Signs for Comparison

The signs that are used to compare fractions are $<$, $>$, and $=$.

$<$ means **is less than**

$>$ means **is greater than**

$=$ means **is equal to**

You can use models or pictures to compare two fractions as long as the wholes containing the fractional parts are both the same size.

The fraction strips show that $\frac{1}{2} > \frac{1}{4}$ because the length of the strip that represents $\frac{1}{2}$ is longer than the strip that represents $\frac{1}{4}$.

When the wholes are different sizes, you cannot compare the fractions.

Example

The models below show a small circle divided into halves and a large circle divided into fourths. The small circle has $\frac{1}{2}$ shaded and the large circle has $\frac{1}{4}$ shaded.

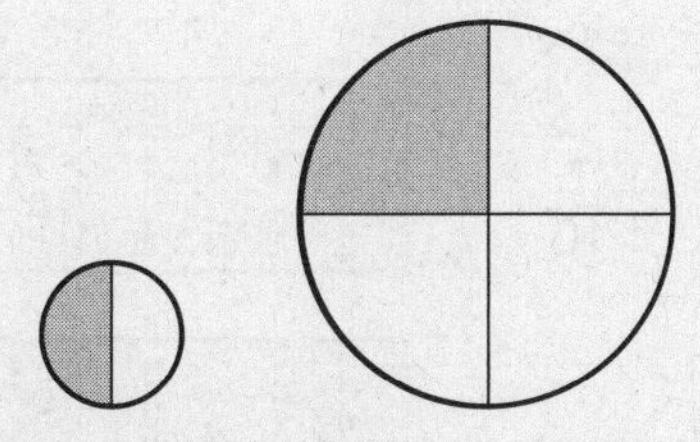

Notice that $\frac{1}{4}$ of the larger circle is bigger than $\frac{1}{2}$ of the small circle.

Example

The models below show two circles that are the same size. One circle is divided into fifths and the other is divided into halves.

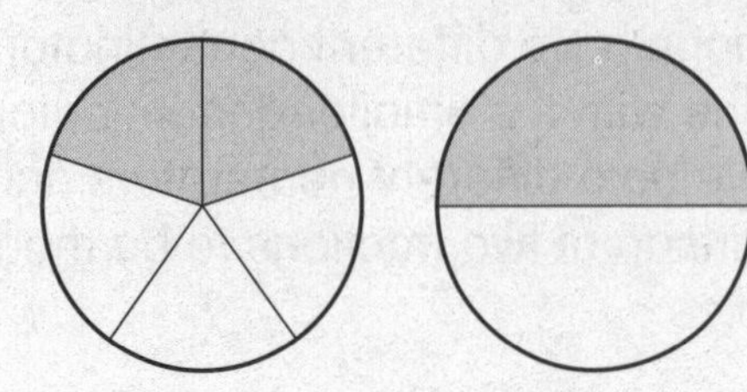

Since the wholes are the same size, you can compare the parts: $\frac{2}{5} < \frac{1}{2}$ and $\frac{1}{2} > \frac{2}{5}$.

You can also compare two fractions when there is no model.

Example

Compare $\frac{3}{4}$ and $\frac{6}{10}$.

Notice that 6 is a multiple of 3. So you can multiply the numerator and denominator of $\frac{3}{4}$ by 2 to make an equivalent fraction with a numerator of 6. Then compare the two fractions by comparing their denominators.

$\frac{3}{4} = \frac{3 \times 2}{4 \times 2} = \frac{6}{8}$ $\qquad \frac{6}{8} \underline{\quad ? \quad} \frac{6}{10}$

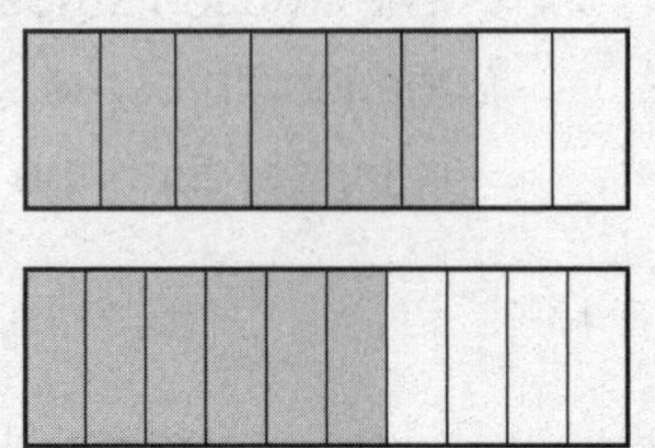

Since eighths are larger than tenths, $\frac{6}{8} > \frac{6}{10}$. You can draw a model to check your comparison.

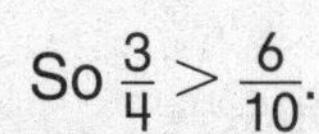

So $\frac{3}{4} > \frac{6}{10}$.

Example

Compare $\frac{3}{5}$ and $\frac{7}{10}$.

Since 10 is a multiple of 5, multiply the numerator and denominator of $\frac{2}{5}$ by 2 to make an equivalent fraction with a denominator of 10. Then compare.

$\frac{3}{5} = \frac{3 \times 2}{5 \times 2} = \frac{6}{10}$ $\qquad \frac{6}{10} \underline{\quad ? \quad} \frac{7}{10}$

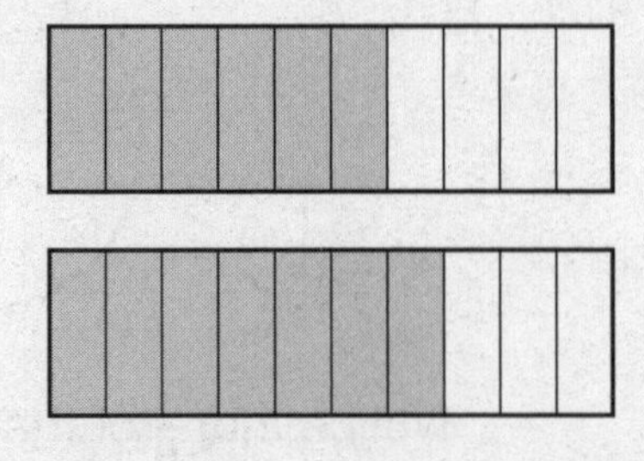

Since all the tenths are the same size, $\frac{6}{10} < \frac{7}{10}$.
You can draw a model to check your comparison.

So $\frac{3}{5} < \frac{7}{10}$.

CCSS: 4.NF.2

Another way to compare fractions is to use benchmarks, such as 0, $\frac{1}{2}$, and 1.

Example

Use $\frac{1}{2}$ as a benchmark to help you compare $\frac{3}{8}$ and $\frac{7}{10}$.

To compare these fractions using the benchmark fraction $\frac{1}{2}$, divide each denominator by 2 and compare the result to the numerator of the fraction. Since $8 \div 2 = 4$ and 3 is less than 4, $\frac{3}{8}$ is less than $\frac{1}{2}$. Since $10 \div 2 = 5$ and 7 is greater than 5, $\frac{7}{10}$ is greater than $\frac{1}{2}$.

So $\frac{3}{8}$ is less than $\frac{7}{10}$. Use the fraction strips on page 130 to check.

If the denominator of a fraction is not a multiple of 2, you can multiply both terms by 2 to make an equivalent fraction before using the benchmark fraction $\frac{1}{2}$.

Example

Use $\frac{1}{2}$ as a benchmark to help you compare $\frac{3}{5}$ and $\frac{5}{12}$.

Multiply both terms of $\frac{3}{5}$ by 2 to make an equivalent fraction with an even denominator.

$$\frac{3}{5} = \frac{3 \times 2}{5 \times 2} = \frac{6}{10}$$

Compare $\frac{6}{10}$ and $\frac{5}{12}$. Since $10 \div 2 = 5$ and 6 is greater than 5, $\frac{6}{10}$ is greater than $\frac{1}{2}$.

Since $\frac{6}{10}$ is equivalent to $\frac{3}{5}$, $\frac{3}{5}$ is greater than $\frac{1}{2}$. Since $12 \div 2 = 6$ and 5 is less than 6, $\frac{5}{12}$ is less than $\frac{1}{2}$.

So $\frac{3}{5} > \frac{5}{12}$. Use the fraction strips on page 130 to check.

Sometimes you can use 0 or 1 as a benchmark to help you compare fractions.

Example

Use 1 as a benchmark to help you compare $\frac{9}{10}$ and $\frac{7}{8}$.

The fraction $\frac{9}{10}$ is $\frac{1}{10}$ less than 1. The fraction $\frac{7}{8}$ is $\frac{1}{8}$ less than 1.

Since the numerators of $\frac{1}{10}$ and $\frac{1}{8}$ are the same, you know that $\frac{1}{10} < \frac{1}{8}$.

The difference between 1 and $\frac{9}{10}$ is less than the difference between 1 and $\frac{7}{8}$.

So, $\frac{9}{10}$ is closer to 1, and $\frac{9}{10} > \frac{7}{8}$.

Use the fraction strips on page 130 to check.

Example

Use 0 as a benchmark to help you compare $\frac{2}{5}$ and $\frac{1}{3}$.

Use the fraction strips on page 130 to see which strip, $\frac{2}{5}$ or $\frac{1}{3}$, is shorter. The shorter strip is closer to 0 and shows the smaller fraction.

Since $\frac{1}{3}$ is closer to 0, $\frac{2}{5} > \frac{1}{3}$.

Practice

Directions: For questions 1 through 4, write the fraction represented by the shaded part of each figure. Then use $>$, $<$, or $=$ to compare the fractions.

1. 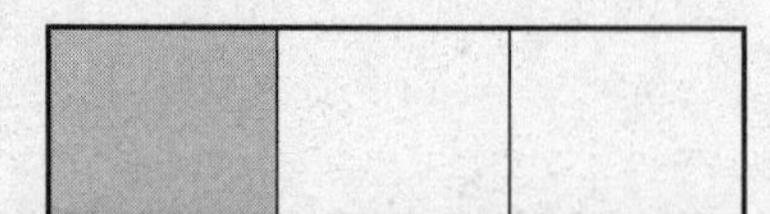

________ ________ ________

2. 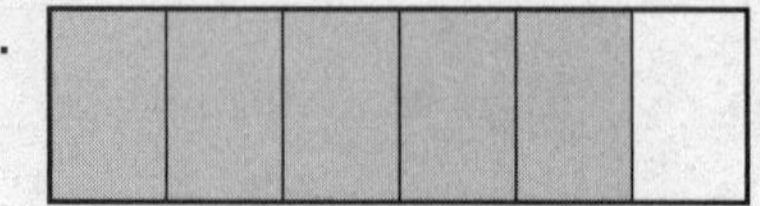 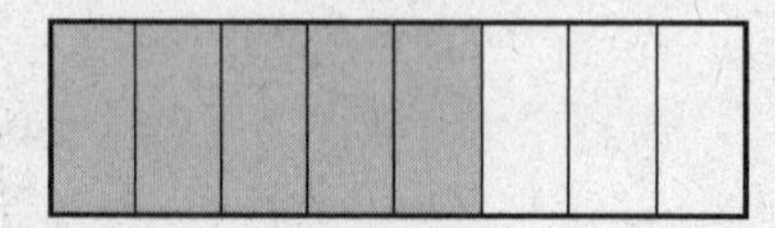

________ ________ ________

CCSS: 4.NF.2

3. 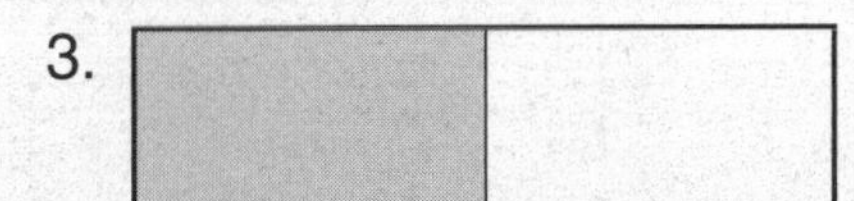 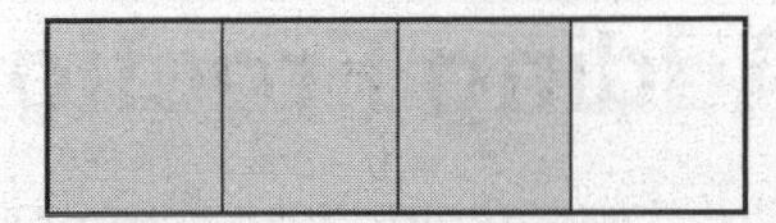

_____ _____ _____

4. 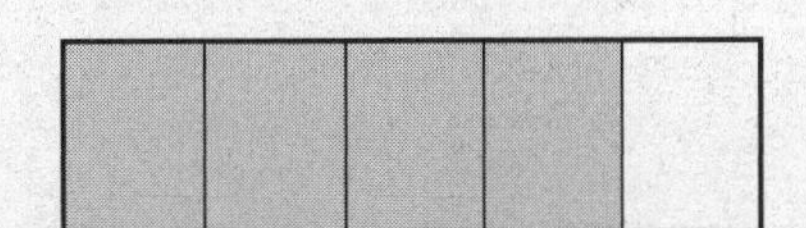

_____ _____ _____

Directions: For questions 5 and 6, compare the fractions by multiplying both terms of one fraction by the same number so its numerator or denominator matches the same term of the other fraction. Then draw a model to check your comparison.

5. Compare $\frac{4}{6}$ and $\frac{8}{10}$.

   $\frac{4}{6}$ _____ $\frac{8}{10}$

6. Compare $\frac{2}{3}$ and $\frac{7}{12}$.

   $\frac{2}{3}$ _____ $\frac{7}{12}$

7. Which fraction is **less than** $\frac{1}{2}$?

   A. $\frac{2}{6}$
   B. $\frac{5}{8}$
   C. $\frac{3}{4}$
   D. $\frac{9}{12}$

8. Which fraction is **greater than** $\frac{1}{2}$?

   A. $\frac{4}{10}$
   B. $\frac{75}{100}$
   C. $\frac{3}{8}$
   D. $\frac{5}{12}$

Lesson 22: Adding Fractions

When you add fractions, you are joining parts that refer to the same whole.
A **unit fraction** has 1 as the numerator.
Any fraction having a numerator greater than 1 can be renamed as one or more addition equations in which each addend is a unit fraction.

Examples

a

$\frac{1}{5}$	$\frac{1}{5}$	$\frac{1}{5}$	=	$\frac{1}{5}$	$\frac{1}{5}$	+	$\frac{1}{5}$

$\frac{3}{5} = \frac{2}{5} + \frac{1}{5}$

b

$\frac{1}{5}$	$\frac{1}{5}$	$\frac{1}{5}$	=	$\frac{1}{5}$	+	$\frac{1}{5}$	+	$\frac{1}{5}$

$\frac{3}{5} = \frac{1}{5} + \frac{1}{5} + \frac{1}{5}$

c

1	1	$\frac{1}{5}$	=	$\frac{1}{5}$	$\frac{1}{5}$	$\frac{1}{5}$	$\frac{1}{5}$	$\frac{1}{5}$	+	$\frac{1}{5}$	$\frac{1}{5}$	$\frac{1}{5}$	$\frac{1}{5}$	$\frac{1}{5}$	+	$\frac{1}{5}$

$2\frac{1}{5} = 1 + 1 + \frac{1}{5} = \frac{5}{5} + \frac{5}{5} + \frac{1}{5}$

To add fractions that have the same denominator, add the numerators and write the sum over the denominator.

Example

Add: $\frac{2}{8} + \frac{3}{8}$

$\frac{1}{8}$	$\frac{1}{8}$	$\frac{1}{8}$	$\frac{1}{8}$	$\frac{1}{8}$	$\frac{1}{8}$	$\frac{1}{8}$	$\frac{1}{8}$	+	$\frac{1}{8}$	$\frac{1}{8}$	$\frac{1}{8}$	$\frac{1}{8}$	$\frac{1}{8}$	$\frac{1}{8}$	$\frac{1}{8}$	$\frac{1}{8}$	=	$\frac{1}{8}$	$\frac{1}{8}$	$\frac{1}{8}$	$\frac{1}{8}$	$\frac{1}{8}$	$\frac{1}{8}$	$\frac{1}{8}$	$\frac{1}{8}$

$\frac{2}{8}$ + $\frac{3}{8}$ = $\mathbf{\frac{5}{8}}$

$\frac{2}{8} + \frac{3}{8} = \frac{2+3}{8} = \frac{5}{8}$

CCSS: 4.NF.3.a, 4.NF.3.b, 4.NF.3.c, 4.NF.3.d

Example

Add: $\frac{1}{6} + \frac{3}{6}$

$\frac{1}{6} + \frac{3}{6} = \frac{1 + 3}{6} = \frac{4}{6}$

Example

Andrew completed $\frac{3}{8}$ of his homework before dinner. He completed $\frac{4}{8}$ more of his homework after dinner. How much of his homework did Andrew complete?

Add $\frac{3}{8}$ and $\frac{4}{8}$.

$\frac{3}{8} + \frac{4}{8} = \frac{3 + 4}{8} = \frac{7}{8}$

Andrew completed $\frac{7}{8}$ of his homework.

You can add mixed numbers by changing them to improper fractions. Then you can change the sum to a mixed number.

Example

Add: $2\frac{3}{8} + 1\frac{7}{8}$

Change each mixed number to an improper fraction.

$2\frac{3}{8} = \frac{(2 \times 8) + 3}{8} = \frac{19}{8}$

$1\frac{7}{8} = \frac{(1 \times 8) + 7}{8} = \frac{15}{8}$

Find the sum of the improper fractions.

$\frac{19}{8} + \frac{15}{8} = \frac{19 + 15}{8} = \frac{34}{8}$

Change the sum to a mixed number.

$\frac{34}{8} = 34 \div 8 = 4\frac{2}{8}$

So $2\frac{3}{8} + 1\frac{7}{8} = 4\frac{2}{8}$

You can use addition properties to add mixed numbers.

Examples

Add: $1\frac{1}{5} + 6\frac{3}{5} + \frac{2}{5}$

Change each mixed number to an improper fraction.

$1\frac{1}{5} = \frac{(1 \times 5) + 1}{5} = \frac{6}{5}$ $6\frac{3}{5} = \frac{(6 \times 5) + 3}{5} = \frac{33}{5}$

Use the commutative property.

$$\frac{6}{5} + \frac{33}{5} + \frac{2}{5} = \frac{6}{5} + \frac{2}{5} + \frac{33}{5}$$
$$= \frac{6 + 2}{5} + \frac{33}{5}$$
$$= \frac{8}{5} + \frac{33}{5}$$
$$= \frac{8 + 33}{5} = \frac{41}{5}$$

Use the associative property.

$$\left(\frac{6}{5} + \frac{33}{5}\right) + \frac{2}{5} = \frac{6}{5} + \left(\frac{33}{5} + \frac{2}{5}\right)$$
$$= \frac{6}{5} + \frac{33 + 2}{5}$$
$$= \frac{6}{5} + \frac{35}{5}$$
$$= \frac{6 + 35}{5} = \frac{41}{5}$$

Change the sum to a mixed number.

$\frac{41}{5} = 41 \div 5 = 8\frac{1}{5}$

So, $1\frac{1}{5} + 6\frac{3}{5} + \frac{2}{5} = 8\frac{1}{5}$.

You can use an equation to represent the information in a word problem. Use a variable such as *n* to represent the unknown number. Then solve the problem.

Example

Tina and Britney shared a pizza. Tina ate $\frac{3}{8}$ of the pizza. Britney ate $\frac{4}{8}$ of the pizza. How much of the pizza did Tina and Britney eat altogether?

The problem is asking for the amount of pizza eaten by both girls.
The given information is the amount eaten by each girl.
You will need to add to find the total amount of pizza eaten.

Write an addition equation that represents the information in the problem.

$\frac{3}{8} + \frac{4}{8} =$ total amount of pizza eaten $= n$

To solve the problem, add the numerators and write the sum over the denominator.

$\frac{3}{8} + \frac{4}{8} = \frac{3 + 4}{8} = \frac{7}{8}$

Tina and Britney ate $\frac{7}{8}$ of the pizza altogether.

CCSS: 4.NF.3.a, 4.NF.3.b, 4.NF.3.c, 4.NF.3.d

Practice

Directions: For questions 1 through 12, add.

1. $\frac{1}{10} + \frac{8}{10} =$ ____________

2. $\frac{3}{8} + \frac{1}{8} =$ ____________

3. $\frac{2}{5} + \frac{2}{5} =$ ____________

4. $\frac{1}{3} + \frac{1}{3} =$ ____________

5. $\frac{1}{4} + \frac{2}{4} =$ ____________

6. $\frac{2}{6} + \frac{3}{6} =$ ____________

7. $3\frac{1}{5} + 2\frac{2}{5} =$ ____________

8. $1\frac{5}{8} + 3\frac{2}{8} =$ ____________

9. $4\frac{1}{3} + 6\frac{1}{3} =$ ____________

10. $2\frac{1}{6} + \frac{1}{6} =$ ____________

11. $7\frac{5}{10} + 2\frac{3}{10} =$ ____________

12. $9\frac{1}{2} + 5\frac{1}{2} =$ ____________

Directions: For questions 13 and 14, rewrite the problem as the sum of unit fractions. Then find the sum.

13. $\frac{3}{10} + \frac{2}{10} =$ ______________ + ______________ = ____________

14. $\frac{4}{8} + \frac{1}{8} =$ ______________ + ______________ = ____________

15. $\left(1\frac{3}{10} + 2\frac{4}{10}\right) + 1\frac{6}{10} = ?$

 A. $1\frac{3}{10} + \left(4\frac{2}{10} + 1\frac{6}{10}\right)$

 B. $1\frac{3}{10} + \left(2\frac{4}{10} + 1\frac{6}{10}\right)$

 C. $1\frac{3}{10} + \left(4\frac{2}{10} + 6\frac{1}{10}\right)$

 D. $3\frac{1}{10} + \left(2\frac{4}{10} + 1\frac{6}{10}\right)$

16. $5\frac{2}{3} + 1\frac{1}{3} =$

 A. $1\frac{1}{3} + \frac{2}{3}$

 B. $1\frac{1}{3} + 5\frac{1}{3}$

 C. $1\frac{1}{3} + 5\frac{2}{3}$

 D. $1\frac{2}{3} + 5\frac{5}{3}$

Solve each problem.

17. Rashid hiked $\frac{1}{8}$ of a mile before lunch and $\frac{3}{8}$ of a mile after lunch. How far did Rashid hike altogether?

    ____

18. Bradley spent $1\frac{1}{6}$ hours raking leaves and $1\frac{2}{6}$ hours mowing the lawn. How long did Bradley rake and mow altogether?

    ____

19. Jenna read $\frac{4}{12}$ of her book on Saturday. She read another $\frac{3}{12}$ of her book on Sunday. How much of the book has Jenna read so far?

    ____

20. It rained $4\frac{3}{5}$ inches in March and $5\frac{4}{5}$ inches in April. How much did it rain in both months?

    ____

    Explain how you got your answer.

    ____

    ____

    ____

CCSS: 4.NF.3.a, 4.NF.3.b, 4.NF.3.c, 4.NF.3.d

Lesson 23: Subtracting Fractions

When you subtract fractions you are removing parts of a whole.

To subtract fractions with like denominators, subtract the numerators and write the difference over the denominator.

Example

Subtract: $\frac{6}{6} - \frac{4}{6}$

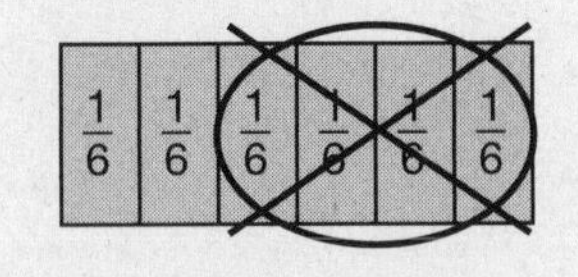

$$\frac{6}{6} - \frac{4}{6} = \frac{2}{6}$$

Example

Subtract: $\frac{4}{5} - \frac{1}{5}$

$$\frac{4}{5} - \frac{1}{5} = \frac{4-1}{5} = \frac{3}{5}$$

Example

Amanda bought $\frac{7}{8}$ of a foot of ribbon. She uses $\frac{3}{8}$ of a foot of ribbon for a school project. How much ribbon does Amanda have left?

Subtract $\frac{3}{8}$ from $\frac{7}{8}$.

$$\frac{7}{8} - \frac{3}{8} = \frac{7-3}{8} = \frac{4}{8}$$

Amanda has $\frac{4}{8}$ of a foot of ribbon left.

Tip: When you subtract fractions that have the same denominator, you are really subtracting the number of unit fractions shown by the numerator of the subtrahend from the number of unit fractions shown by the numerator of the minuend. For example, $\frac{4}{5} - \frac{3}{5}$ is $\frac{1}{5} + \frac{1}{5} + \frac{1}{5} + \frac{1}{5}$ minus $\frac{1}{5} + \frac{1}{5} + \frac{1}{5}$, a difference of $\frac{1}{5}$.

You can subtract mixed numbers by changing them to improper fractions. Then you can change the difference to a mixed number.

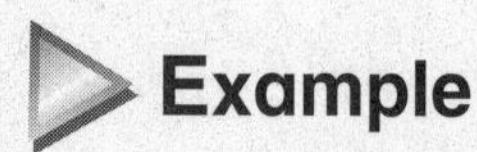

Example

Subtract: $4\frac{1}{6} - 2\frac{5}{6}$

Change each mixed number to an improper fraction.

$$4\frac{1}{6} = \frac{(4 \times 6) + 1}{6} = \frac{25}{6}$$

$$2\frac{5}{6} = \frac{(2 \times 6) + 5}{6} = \frac{17}{6}$$

Find the difference of the improper fractions.

$$\frac{25}{6} - \frac{17}{6} = \frac{25 - 17}{6} = \frac{8}{6}$$

Change the difference to a mixed number.

$$\frac{8}{6} = 8 \div 6 = 1\frac{2}{6}$$

So $4\frac{1}{6} - 2\frac{5}{6} = 1\frac{2}{6}$.

Because addition and subtraction are "opposites," you can use addition to check your answer to a subtraction problem. When you add the difference and the number that you subtracted, your sum should be the number that you subtracted from.

Example

The difference in the example above is $1\frac{2}{6}$. The number you subtracted is $2\frac{5}{6}$.

$$1\frac{2}{6} = \frac{8}{6}$$

$$2\frac{5}{6} = \frac{17}{6}$$

$$\frac{8}{6} + \frac{17}{6} = \frac{8 + 17}{6} = \frac{25}{6}$$

$$\frac{25}{6} = 25 \div 6 = 4\frac{1}{6}$$

So $1\frac{2}{6} + 2\frac{5}{6} =$ the number you subtracted from

Since the sum is the number you subtracted from, your answer to the subtraction problem is correct.

CCSS: 4.NF.3.a, 4.NF.3.b, 4.NF.3.c, 4.NF.3.d

You can use an equation to represent the information in a word problem. Use a variable such as *n* to represent the unknown number. Then use the appropriate operation to solve the problem.

Example

Eric bought $\frac{7}{10}$ of a pound of turkey and $\frac{3}{10}$ of a pound of ham. How much more turkey than ham did Eric buy?

The problem is asking for the difference between the weights of the meats purchased.

The information given is that Eric bought $\frac{7}{10}$ pound of turkey and $\frac{3}{10}$ pound of ham. You will need to subtract to find the difference.

Write a subtraction equation that represents the information in the problem.

$$\frac{7}{10} - \frac{3}{10} = \text{difference between the weights of the meats} = n$$

To solve the problem, subtract the numerators and write the difference over the denominator.

$$\frac{7}{10} - \frac{3}{10} = \frac{7-3}{10} = \frac{4}{10}$$

So $\frac{7}{10} - \frac{3}{10} = \frac{4}{10}$.

Eric bought $\frac{4}{10}$ of a pound more turkey than ham.

Practice

Directions: For questions 1 and 2, rewrite the problem as the difference between unit fractions. Then find the difference.

1. $\frac{4}{5} - \frac{2}{5} =$ ________________ − ________________ = ____________

2. $\frac{3}{12} - \frac{2}{12} =$ ________________ − ________________ = ____________

Directions: For questions 3 through 20, subtract.

3. $\frac{3}{4} - \frac{1}{4} =$ ____________

4. $\frac{7}{10} - \frac{5}{10} =$ ____________

5. $\frac{5}{6} - \frac{1}{6} =$ ____________

6. $\frac{2}{3} - \frac{1}{3} =$ ____________

7. $\frac{3}{8} - \frac{2}{8} =$ ____________

8. $\frac{4}{5} - \frac{2}{5} =$ ____________

9. $\frac{3}{5} - \frac{2}{5} =$ ____________

10. $\frac{7}{8} - \frac{3}{8} =$ ____________

11. $\frac{9}{10} - \frac{4}{10} =$ ____________

12. $5\frac{1}{4} - 1\frac{3}{4} =$ ____________

13. $3\frac{5}{8} - 1\frac{2}{8} =$ ____________

14. $2\frac{9}{10} - 1\frac{2}{10} =$ ____________

15. $6\frac{1}{3} - 4\frac{2}{3} =$ ____________

16. $4\frac{1}{5} - 3\frac{4}{5} =$ ____________

17. $7\frac{5}{6} - 3\frac{1}{6} =$ ____________

18. $3\frac{5}{10} - 2\frac{7}{10} =$ ____________

19. $5\frac{3}{5} - 4\frac{4}{5} =$ ____________

20. $3\frac{5}{6} - 1\frac{3}{6} =$ ____________

Directions: For questions 21 and 22, choose the addition problem that can be used to check the answer to the subtraction problem.

21. $\frac{4}{6} - \frac{2}{6} = \frac{2}{6}$

A. $\frac{2}{6} + \frac{4}{6} = \frac{6}{6}$

B. $\frac{4}{6} + \frac{4}{6} = \frac{8}{6}$

C. $\frac{2}{6} + \frac{2}{6} = \frac{4}{6}$

D. $\frac{1}{6} + \frac{3}{6} = \frac{4}{6}$

22. $6\frac{1}{3} - 2\frac{2}{3} = 3\frac{2}{3}$

A. $3\frac{2}{3} + 2\frac{2}{3} = 6\frac{1}{3}$

B. $\frac{1}{3} + 6 = 6\frac{1}{3}$

C. $3\frac{1}{3} + 3 = 6\frac{1}{3}$

D. $3\frac{2}{3} + 3\frac{1}{3} = 7$

CCSS: 4.NF.3.a, 4.NF.3.b, 4.NF.3.c, 4.NF.3.d

Solve each problem.

23. Carla's gas tank was $\frac{2}{3}$ full. After she drove to Rapid Falls, her gas tank was $\frac{1}{3}$ full. What fraction of a tank of gas did Carla use to drive to Rapid Falls?

____

24. Ahmad plans to run $3\frac{1}{5}$ miles. He ran $1\frac{4}{5}$ miles before he stopped for a break. How many miles does Ahmad have left to run?

____

25. Christopher was at the beach collecting seashells. He filled up his bucket $\frac{7}{8}$ of the way with seashells. He gave $\frac{3}{8}$ of the seashells to his brother. How full was his bucket with seashells then?

____

26. Tanesha did some babysitting for $4\frac{1}{6}$ hours on Friday and for $2\frac{5}{6}$ hours on Saturday. How much more time did Tanesha spend babysitting on Friday than on Saturday?

____

Use addition to check your answer. Explain what you did.

____

____

____

Lesson 24: Multiplying Fractions and Whole Numbers

Any given fraction with a numerator greater than 1 can be thought of as a multiple of a unit fraction having the same denominator as the given fraction.

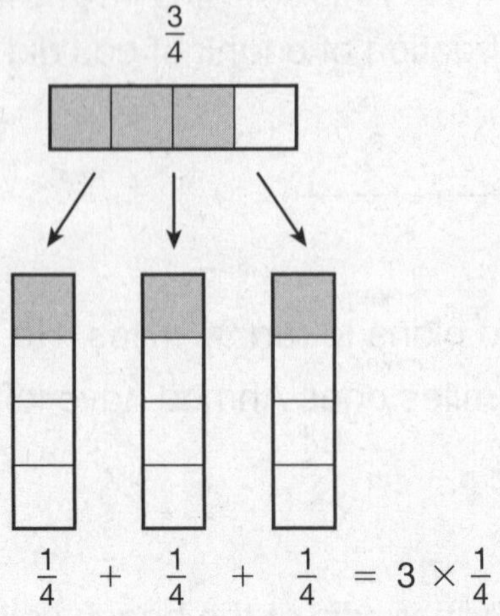

For example, $\frac{3}{4}$ is a multiple of $\frac{1}{4}$

because $\frac{3}{4} = \frac{1}{4} + \frac{1}{4} + \frac{1}{4}$, and

$\frac{1}{4} + \frac{1}{4} + \frac{1}{4}$ is equivalent to $3 \times \frac{1}{4}$.

So $\frac{3}{4} = 3 \times \frac{1}{4}$.

You have already added fractions and multiplied with whole numbers. You can use what you know to multiply a fraction and a whole number. Since multiplication is repeated addition, just use the whole number to see how many times the fraction will be added.

Example

Multiply by using repeated addition: $4 \times \frac{2}{3}$

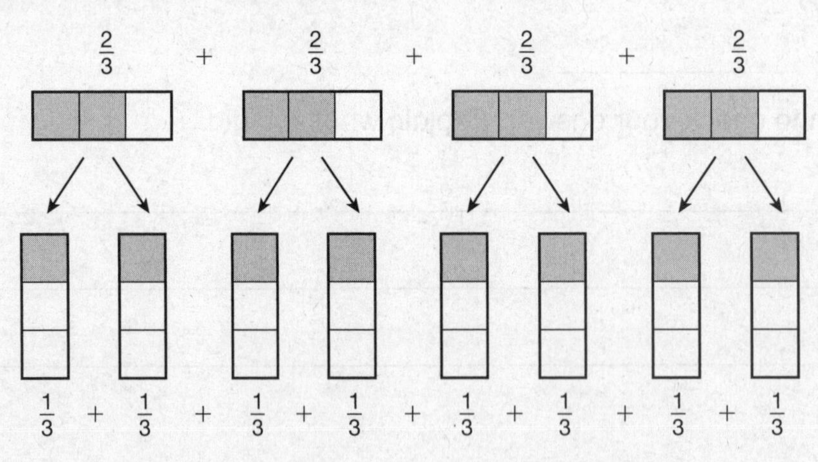

CCSS: 4.NF.4a, 4.NF.4b, 4.NF.4c

The model on page 152 shows that each $\frac{2}{3}$ is equivalent to $\frac{1}{3} + \frac{1}{3}$. Add the unit fractions of the model to find the total number of thirds. Write the total number of thirds as the numerator of the sum. The denominator of the sum will still be 3, because this shows how many equal parts make up each shape in the model.

$\frac{2}{3} + \frac{2}{3} + \frac{2}{3} + \frac{2}{3} = \frac{8}{3}$

So $4 \times \frac{2}{3} = \frac{8}{3}$ because 4 groups of $\frac{2}{3}$ each is another way to represent the same number as $\frac{1}{3} + \frac{1}{3} + \frac{1}{3} + \frac{1}{3} + \frac{1}{3} + \frac{1}{3} + \frac{1}{3} + \frac{1}{3}$, or $\frac{8}{3}$.

So $4 \times \frac{2}{3} = 8 \times \frac{1}{3}$ or $\frac{8}{3}$.

The short way to multiply a fraction and a whole number is to multiply the numerator of the fraction by the whole number and to write the product over the denominator of the fraction.

Example

Multiply: $4 \times \frac{3}{4}$

The whole number is 4, and the numerator of the fraction is 3.

Since $4 \times 3 = 12$, the numerator of the product will be 12.

The denominator of the fraction is 4, so the denominator of the product will be 4.

So $4 \times \frac{3}{4} = \frac{4 \times 3}{4} = \frac{12}{4}$.

Example

Multiply: $8 \times \frac{3}{5}$

The whole number is 8 and the numerator of the fraction is 3.

Since $8 \times 3 = 24$, the numerator of the product will be 24.

The denominator of the fraction is 5, so the denominator of the product will be 5.

So $8 \times \frac{3}{5} = \frac{8 \times 3}{5} = \frac{24}{5}$.

You can use an equation to represent the information in a word problem. Use a variable such as *n* to represent the unknown number. Then solve the problem.

Example

Each team member in a relay race will run $\frac{3}{4}$ of a mile, and there are 5 members on the team. How many miles will the team members run in all? Between what two whole numbers does your answer lie?

Write an equation that represents the given information.

$$5 \times \frac{3}{4} = \text{number of miles run} = n$$

To solve the problem, multiply the numerator of the fraction by the whole number. Then write the product over the denominator of the fraction.

$$5 \times \frac{3}{4} = \frac{5 \times 3}{4} = \frac{15}{4}$$

The team members will run a total of $\frac{15}{4}$ miles.

You can tell that the fraction $\frac{15}{4}$ is not a whole number because the numerator is not a multiple of 4. The multiple of 4 that is closest to 15 is 16, and $16 \div 4 = 4$. Since 15 is less than 16, $\frac{15}{4}$ is less than 4. The closest multiple of 4 that is less than 16 is 12. Since $12 \div 4 = 3$ and $16 \div 4 = 4$, the answer is between 3 and 4.

Check your answer by drawing a model for $5 \times \frac{3}{4}$ and using repeated addition.

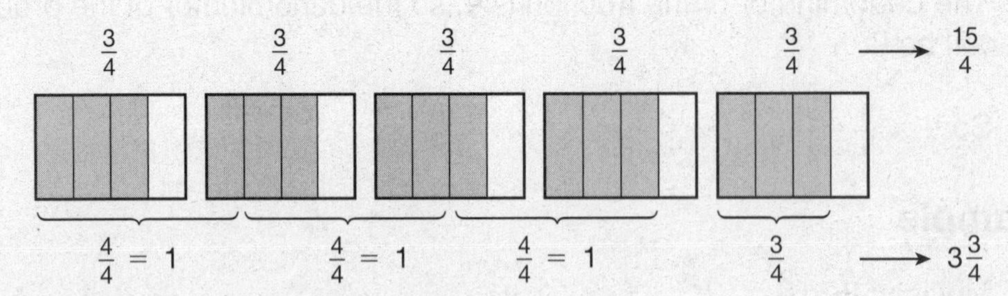

So $5 \times \frac{3}{4} = \frac{3}{4} + \frac{3}{4} + \frac{3}{4} + \frac{3}{4} + \frac{3}{4} = \frac{15}{4}$

and $5 \times \frac{3}{4} = 1 + 1 + 1 + \frac{3}{4} = 3\frac{3}{4}$.

CCSS: 4.NF.4a, 4.NF.4b, 4.NF.4c

Practice

Directions: For questions 1 through 12, multiply.

1. $4 \times \frac{1}{2} =$ __________

2. $\frac{7}{12} \times 3 =$ __________

3. $\frac{1}{3} \times 8 =$ __________

4. $5 \times \frac{5}{6} =$ __________

5. $\frac{2}{5} \times 10 =$ __________

6. $4 \times \frac{7}{10} =$ __________

7. $\frac{3}{8} \times 15 =$ __________

8. $6 \times \frac{1}{6} =$ __________

9. $\frac{1}{2} \times 28 =$ __________

10. $12 \times \frac{2}{3} =$ __________

11. $\frac{3}{20} \times 3 =$ __________

12. $11 \times \frac{2}{7} =$ __________

Directions: For questions 13 and 14, rewrite the problem as the sum of unit fractions and then as the product of a whole number and a unit fraction. Then find the product.

13. $2 \times \frac{3}{8} =$ __________ = __________ × __________ = __________

14. $4 \times \frac{2}{5} =$ __________ = __________ × __________ = __________

15. $\frac{4}{6} = 4 \times$ ____?____

   A. $\frac{1}{4}$

   B. $\frac{1}{5}$

   C. $\frac{1}{6}$

   D. 6

16. ____?____ $= 7 \times \frac{1}{8}$

   A. $\frac{7}{8}$

   B. $\frac{8}{7}$

   C. $7\frac{1}{8}$

   D. 8

Solve each problem.

17. The track at Brownsville Elementary School is $\frac{9}{10}$ of a mile long. Petra ran 3 laps of the track. How many miles did Petra run? Between what two whole numbers does your answer lie?

    ____

18. Mike made 7 cushion covers. He used $\frac{7}{8}$ of a yard of fabric for each cover. How many yards of fabric did Mike use? Between what two whole numbers does your answer lie?

    ____

19. If one burger uses $\frac{1}{4}$ of a pound of turkey, how many pounds of turkey do 10 burgers use? Between what two whole numbers does your answer lie?

    ____

20. It takes Luis $\frac{5}{6}$ of an hour to build one birdhouse. How many hours does it take Luis to build 4 birdhouses? Between what two whole numbers does your answer lie?

    ____

    Check your answer by drawing a model and using repeated addition.

Lesson 25: Relating Decimals to Fractions

A **decimal** is a number that shows multiples of $\frac{1}{10}$ and $\frac{1}{100}$ by using a **decimal point**.

You have used a place-value table to show the value of each digit in a whole number. You can also use a place-value table to show a fraction that is a multiple of $\frac{1}{10}$ or $\frac{1}{100}$. The pattern in the place-value table is the same for both whole numbers and decimals: The value of each place shown on the table is 10 times greater than the value of the place next to it on the right and $\frac{1}{10}$ of the value of the place next to it on the left.

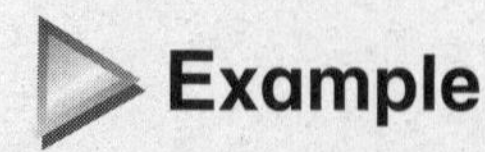

Example

What is the value represented by the numeral shown in the place-value table below?

Hundreds	Tens	Ones	Decimal Point	Tenths	Hundredths
		0	.	2	3

The value of the digit 2 in the tenths place is $\frac{2}{10}$.
You write 0.2 to represent the decimal form of the fraction $\frac{2}{10}$.
In word form, you say or read the decimal 0.2 as *two tenths.*

The value of the digit 3 in the hundredths place is $\frac{3}{100}$.
You write 0.03 to represent the decimal form of the fraction $\frac{3}{100}$.
In word form, you say or read the decimal 0.03 as *three hundredths.*

You write 0.23 to represent the decimal form of the fraction $\frac{23}{100}$, which is equivalent to $0.2 + 0.03$, or $\frac{2}{10} + \frac{3}{100}$.
In word form, you say or read the decimal 0.23 as *twenty-three hundredths.*
Since there are no ones, a zero is written in the ones place.

You can use decimal models to show a part of a whole that is a multiple of $\frac{1}{10}$ or $\frac{1}{100}$.

Example

What decimal is equivalent to the fraction $\frac{3}{10}$?

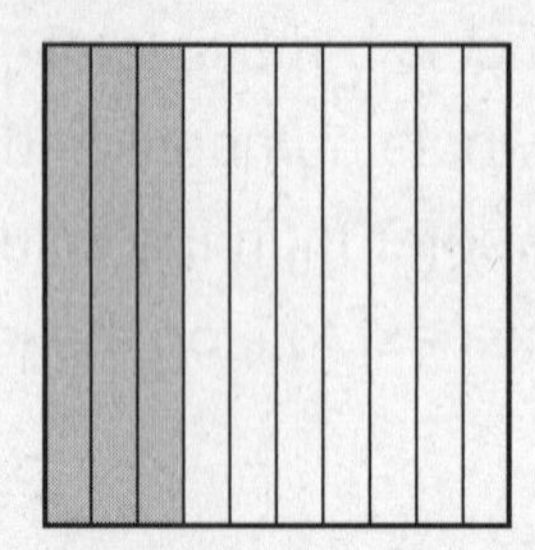

In the model, 3 of 10 equal parts are shaded. The tenths digit will be 3. So the decimal 0.3 is equivalent to the fraction $\frac{3}{10}$.

Example

What decimal is equivalent to the fraction $\frac{5}{100}$?

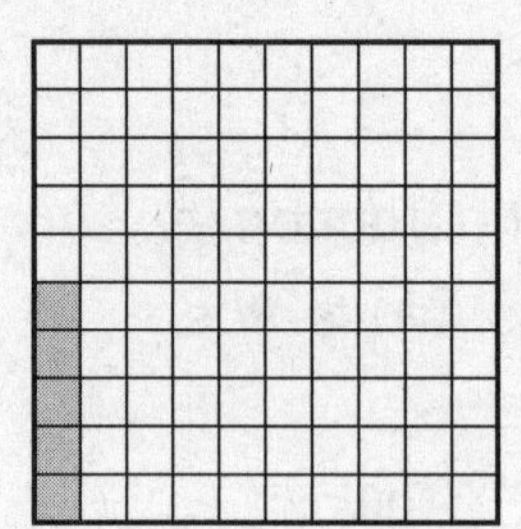

In the model, 5 of 100 equal parts are shaded, so the tenths digit will be 0, and the hundredths digit will be 5. The decimal 0.05 is equal to $\frac{5}{100}$.

Example

What decimal and fraction are shown by the shaded part of the model?

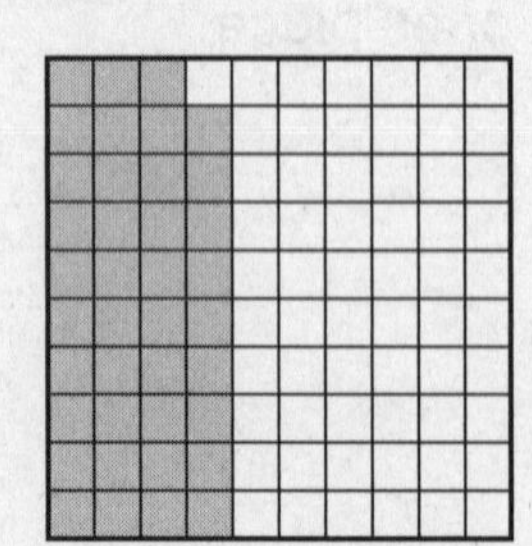

The model shows 100 equal parts with 39 of the parts shaded. So the shaded part of the model stands for 0.39, or $\frac{39}{100}$. The decimal 0.39 is equivalent to $\frac{3}{10} + \frac{9}{100}$.

You can use models to find equivalent decimals.

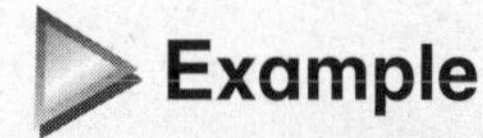

Example

What number does each model stand for?

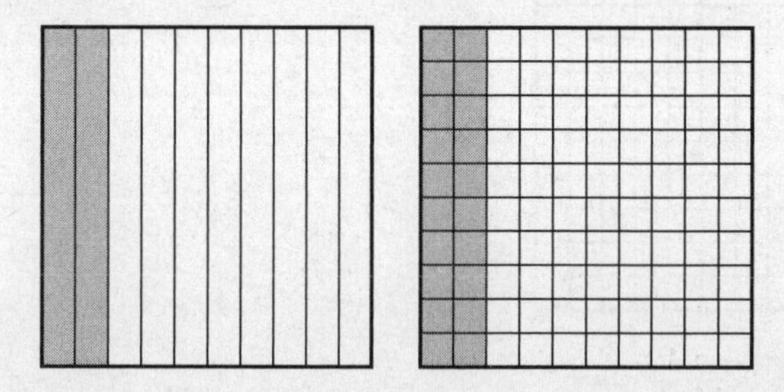

The model on the left shows 2 of 10 equal parts. The model represents the fraction $\frac{2}{10}$, or two tenths. You write 0.2 to represent the decimal two tenths.

The model on the right shows 20 of 100 equal parts. It stands for the fraction $\frac{20}{100}$, or twenty hundredths. You write 0.20 to represent the decimal twenty hundredths.

Each large square represents 1 whole. Since the two wholes are the same size and the shaded regions of the two wholes are the same size, the models show that the decimals 0.2 and 0.20 are equivalent.

Example

You have used a point on a number line to represent the location of a whole number. You can also use a point on a number line to represent the location of a decimal.

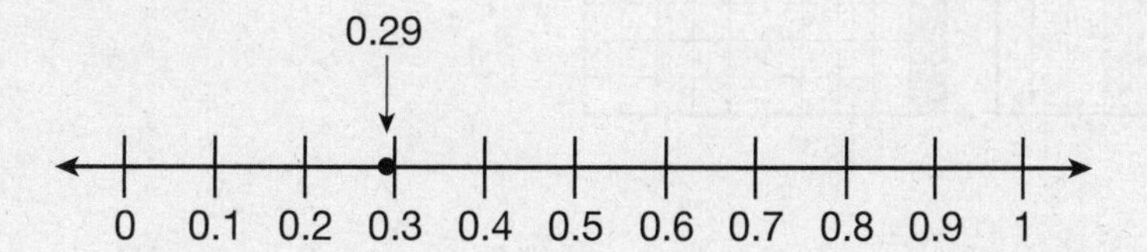

You can model and write decimal numerals for numbers greater than 1.

Example

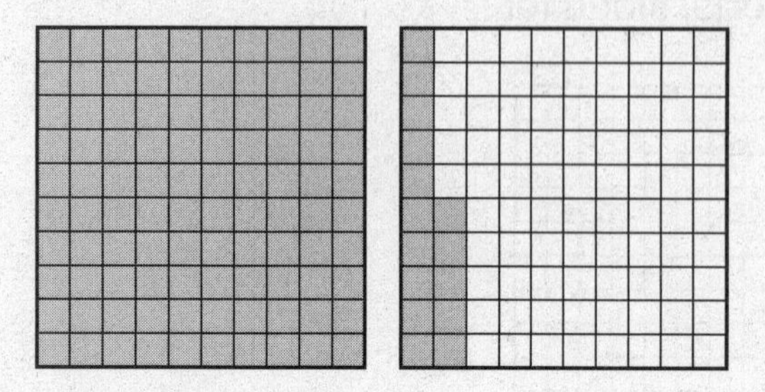

When you say or read the number, you use the word "and" to separate the whole number part from the decimal part. So you say or read 1.15 as *one and fifteen hundredths*.

To add two fractions when one has a denominator of 10 and the other has a denominator of 100, you can express the fraction with the denominator of 10 as an equivalent fraction having a denominator of 100.

Example

What is the sum of $\frac{7}{10}$ and $\frac{3}{100}$?

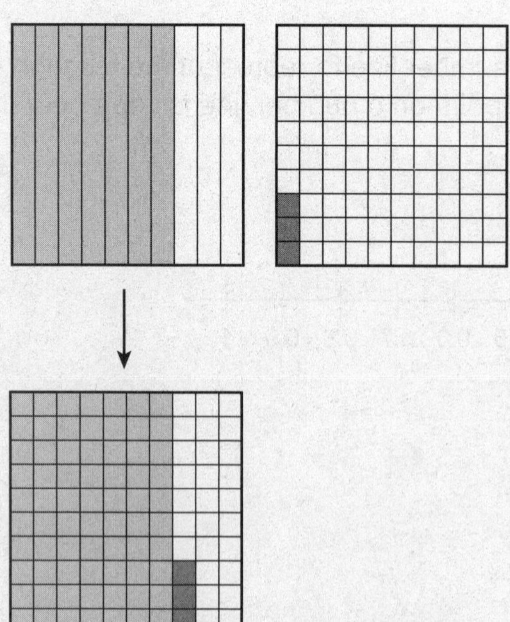

Since $\frac{7}{10}$ is equivalent to $\frac{70}{100}$, $\frac{7}{10} + \frac{3}{100} = \frac{70}{100} + \frac{3}{100} = \frac{73}{100}$.

CCSS: 4.NF.5, 4.NF.6

Practice

Directions: For questions 1 through 8, name the digit in the tenths place.

1. 0.83 ____________
2. 0.40 ____________
3. 0.36 ____________
4. 0.52 ____________
5. 0.25 ____________
6. 0.01 ____________
7. 0.68 ____________
8. 0.73 ____________

Directions: For questions 9 through 16, name the digit in the hundredths place.

9. 0.19 ____________
10. 0.06 ____________
11. 0.43 ____________
12. 0.95 ____________
13. 0.37 ____________
14. 0.50 ____________
15. 0.04 ____________
16. 0.26 ____________

Directions: For questions 17 through 26, write the decimal equivalent for the fraction.

17. $\frac{2}{10}$ ____________
18. $\frac{19}{100}$ ____________
19. $\frac{40}{100}$ ____________
20. $\frac{9}{10}$ ____________
21. $\frac{57}{100}$ ____________
22. $\frac{90}{100}$ ____________
23. $\frac{7}{10}$ ____________
24. $\frac{5}{10}$ ____________
25. $\frac{33}{100}$ ____________
26. $\frac{81}{100}$ ____________

Directions: For questions 27 and 28, write the decimal and word name for the shaded part.

27. 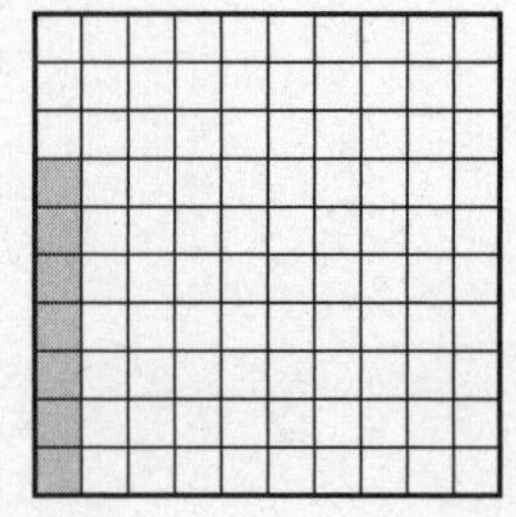

____

28. 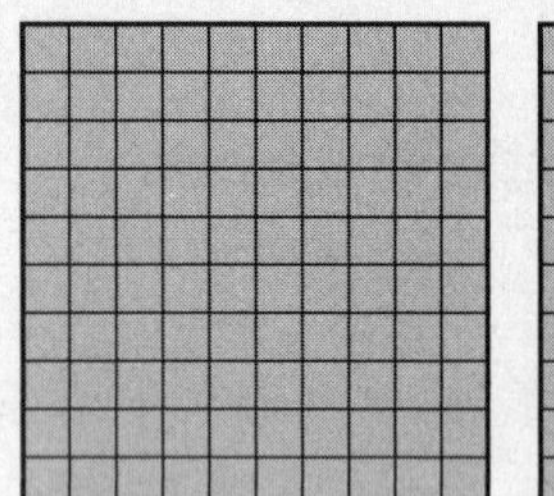 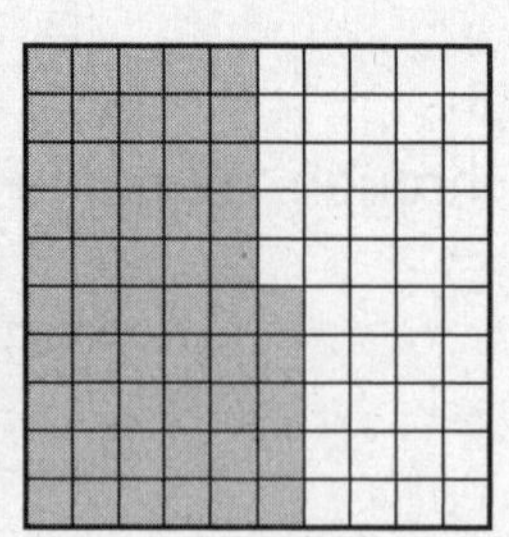

____

29. What is the value of the digit 6 in 0.26?

A. $\frac{6}{10}$

B. 6

C. $\frac{6}{100}$

D. 60

30. What is the value of the digit 3 in 0.39?

A. thirty

B. three

C. three tenths

D. three hundredths

31. What is the value of the digit 4 in 2.41?

A. four tenths

B. four

C. four hundredths

D. forty

32. What is the value of the digit 8 in 3.85?

A. 80

B. $\frac{8}{10}$

C. 8

D. $\frac{8}{100}$

CCSS: 4.NF.5, 4.NF.6

33. Draw and label a point that represents 0.64 on the number line.

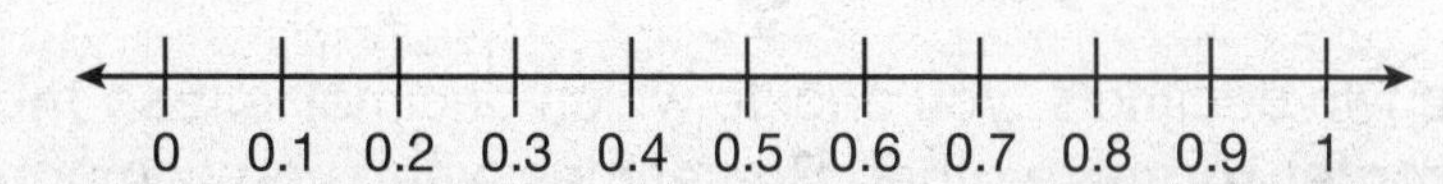

Directions: For questions 34 and 35, find the sum. Express each answer as a fraction.

34. 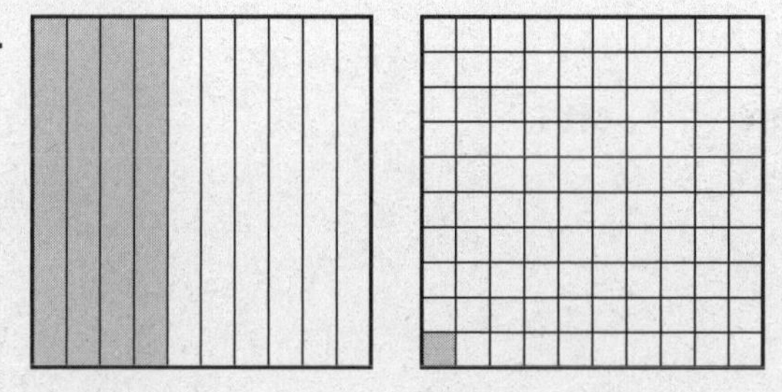

____

Explain how you got your answer.

____

____

____

35. 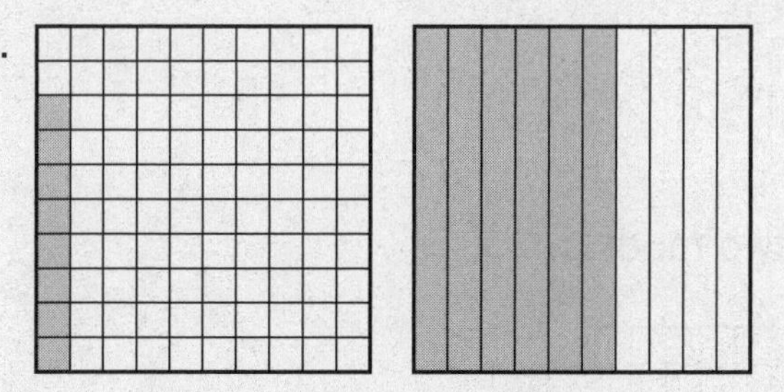

____

Explain how you got your answer.

____

____

____

Lesson 26: Comparing Decimals

When you compare two decimals, you decide which decimal is **less than** the other or which decimal is **greater than** the other. Sometimes when you compare two decimals, you find that the decimals are **equal**.

Signs for Comparison

The signs that are used to compare decimals are $<$, $>$, and $=$.

$<$ means **is less than**

$>$ means **is greater than**

$=$ means **is equal to**

Examples

0.21 is greater than 0.12, so $0.21 > 0.12$.

0.34 is equal to 0.34, so $0.34 = 0.34$.

0.07 is less than 0.7, so $0.07 < 0.7$.

You can use models or pictures to compare two decimals as long as the wholes containing the parts are both the same size.

Example

Compare the shaded parts of these two models.

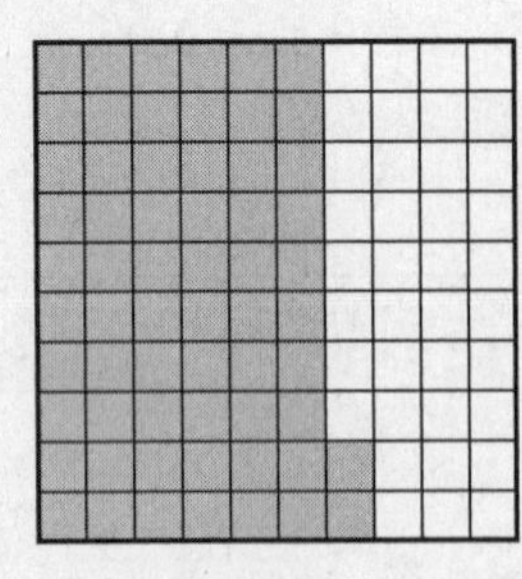

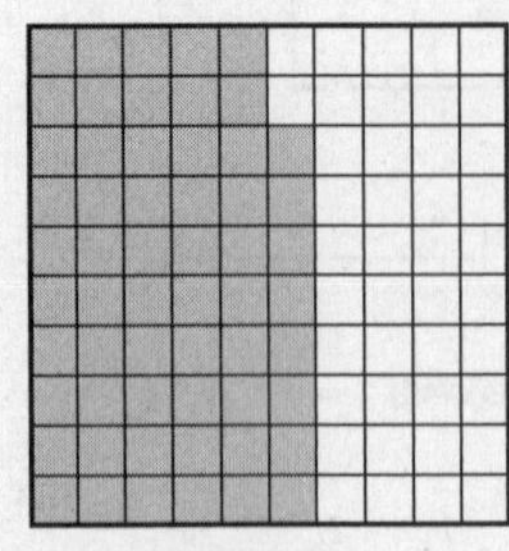

0.62 $>$ 0.58

You can also use place-value tables to compare decimals. Follow the same procedure as you did when comparing whole numbers. Look at the digits in each place-value position from left to right and find the first place-value position in which the digits are different.

Example

Compare the decimals 0.29 and 0.27.

The following place-value table shows that the two decimals have the same digits in the ones and tenths places. The first place in which the digits are different is the hundredths place.

Hundreds	Tens	Ones	Decimal Point	Tenths	Hundredths
		0	.	2	9
		0	.	2	7

The digit 9 represents 9 hundredths, and the digit 7 represents 7 hundredths. Nine hundredths is greater than 7 hundredths.

So 0.29 is greater than 0.27. Write: $0.29 > 0.27$

It is also true that 0.27 is less than 0.29. Write: $0.27 < 0.29$

To check your comparison, shade 27 hundredths of one large square and 29 hundredths of the other. Compare the sizes of the shaded parts.

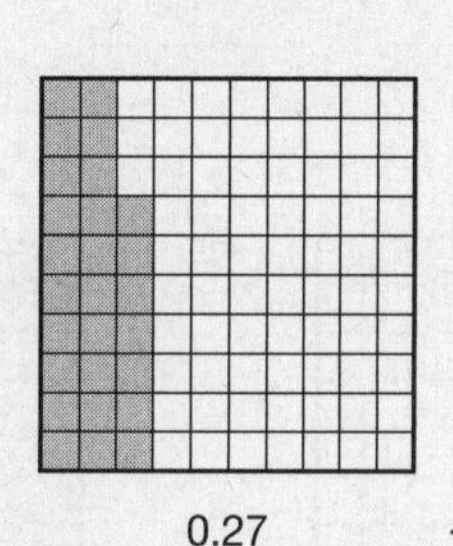
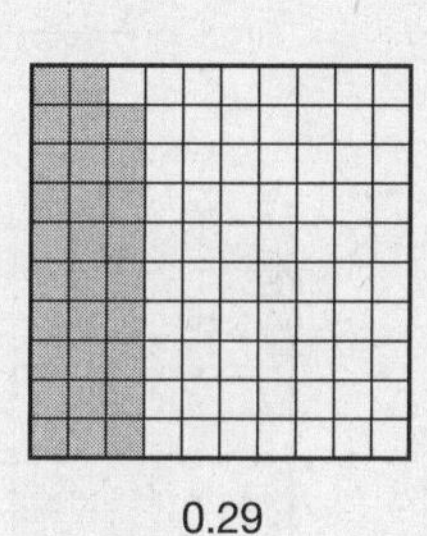

0.27 < 0.29

Example

Compare the numbers 198.43 and 198.51.

Hundreds	Tens	Ones	Decimal Point	Tenths	Hundredths
1	9	8	.	4	3
1	9	8	.	5	1

Compare the hundreds: The hundreds digits are the same.
Compare the tens: The tens digits are the same.
Compare the ones: The ones digits are the same.
Compare the tenths: The digit 4 represents 4 tenths, and the digit 5 represents 5 tenths. Five tenths is greater than 4 tenths.

So 198.51 is greater than 198.43. Write: $198.51 > 198.43$

It is also true that 198.43 is less than 198.51. Write: $198.43 < 198.51$

 Practice

Directions: Under each model for questions 1 through 4, write the decimal represented. Then write <, =, or > to compare the two decimals.

1. 

______ ______ ______

2. 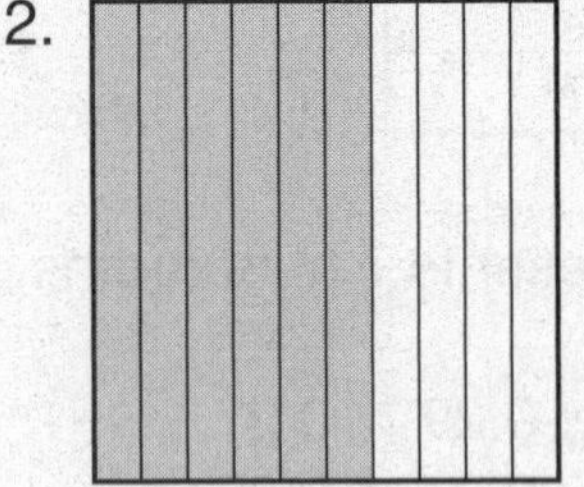 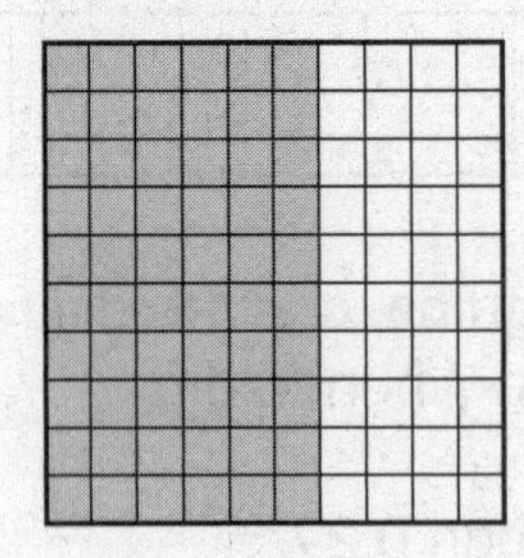

______ ______ ______

3. 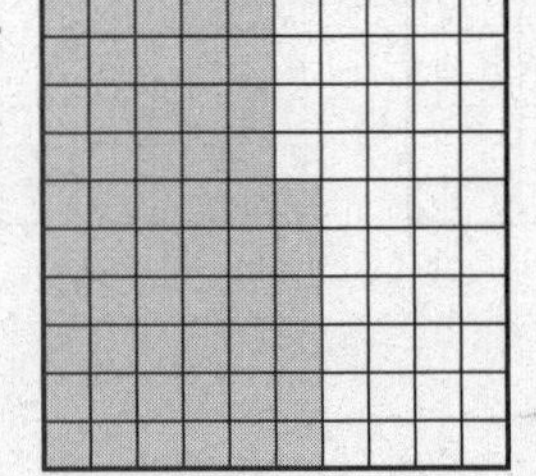 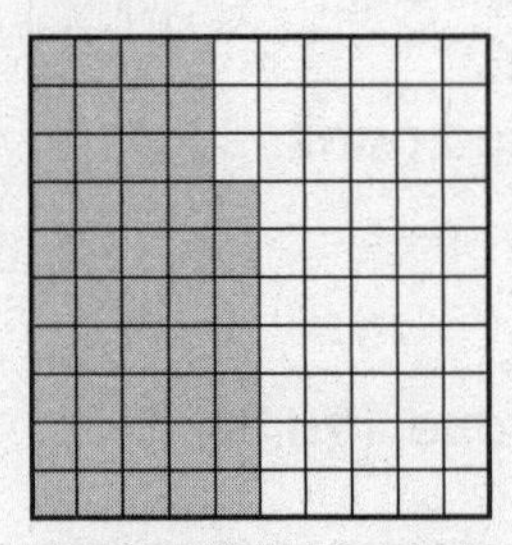

______ ______ ______

4. 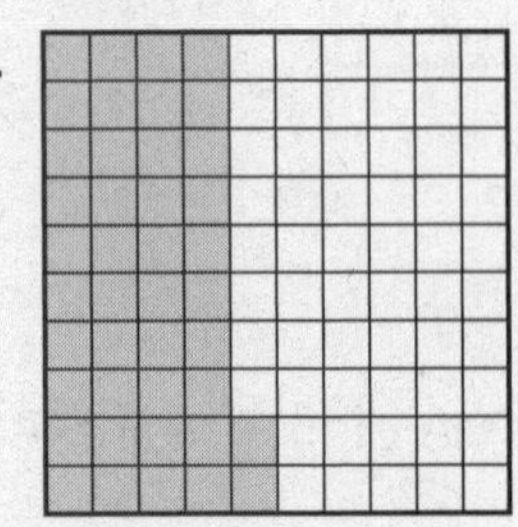 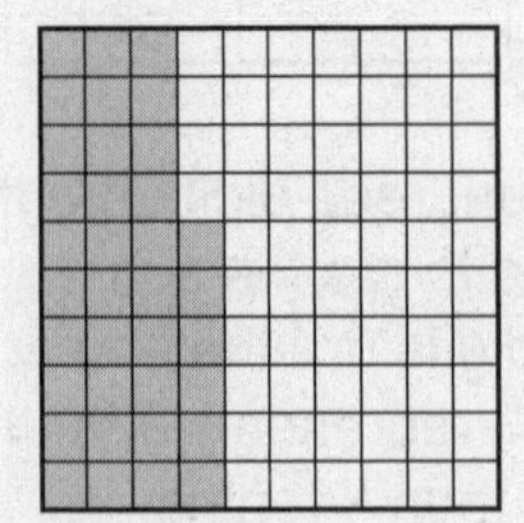

______ ______ ______

CCSS: 4.NF.7

Directions: For questions 5 through 8, shade each model to represent the decimal that is shown below it. Then write $<$, $=$, or $>$ to compare the two decimals.

5. 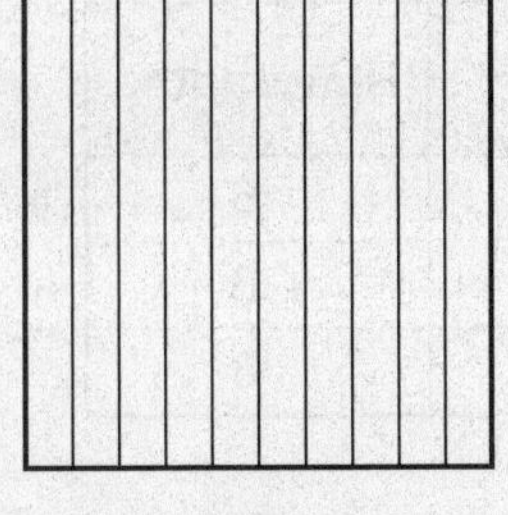 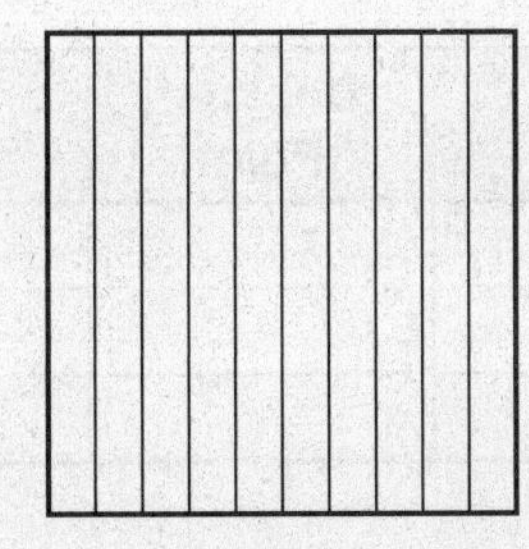

0.4 ______ 0.5

6. 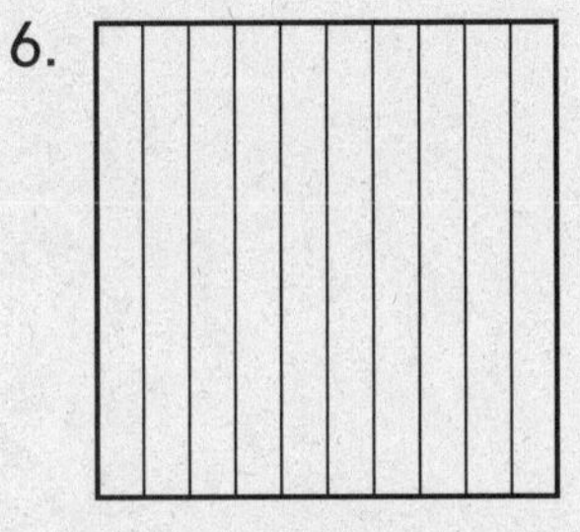 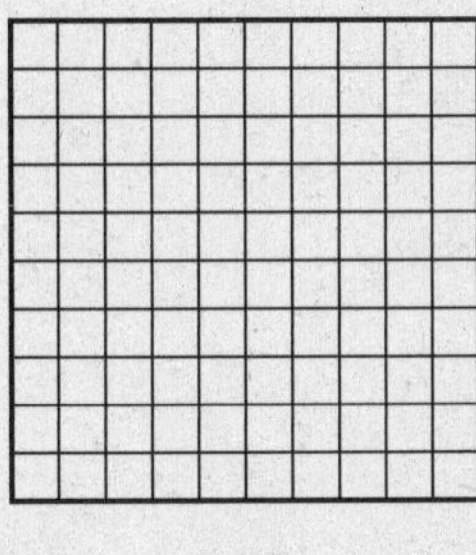

0.8 ______ 0.80

7. 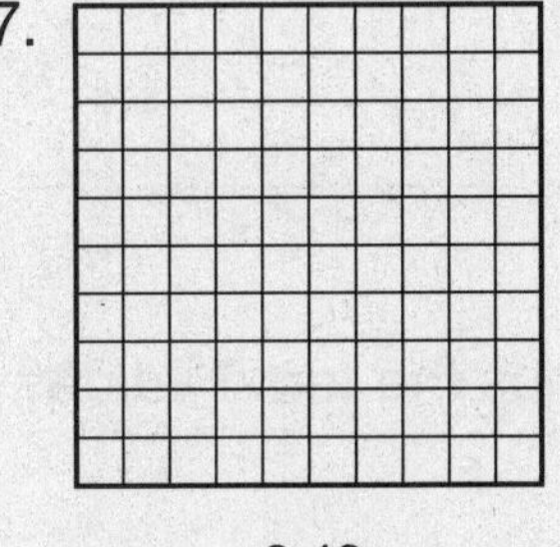 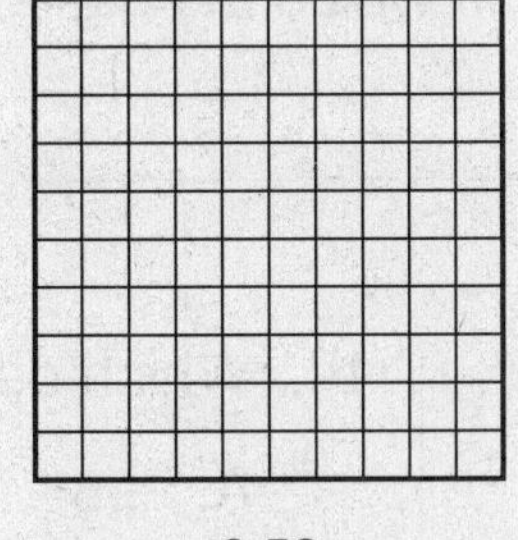

0.48 ______ 0.58

8. 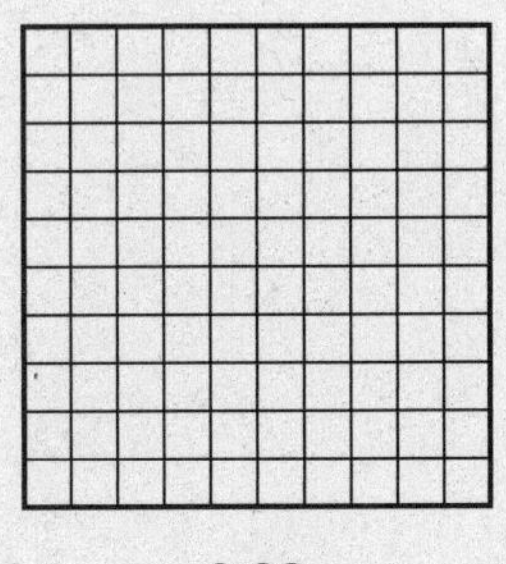 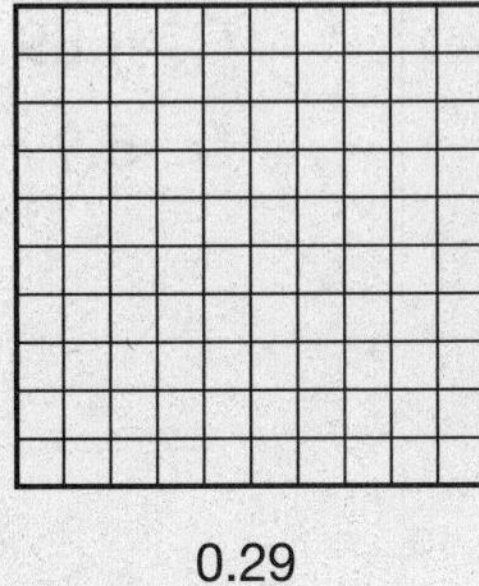

0.32 ______ 0.29

Directions: Use the following table to answer questions 9 through 12. For questions 9 through 11, use <, >, or = to compare the decimals.

Hundreds	Tens	Ones	Decimal Point	Tenths	Hundredths
		0	.	5	6
		0	.	5	8
		0	.	5	3

9. 0.56 ______ 0.53

10. 0.58 ______ 0.56

11. 0.53 ______ 0.58

12. What is the first place-value position in which the digits are different?

____

13. Which decimal has the **greatest** value?

    A. 1.3

    B. 1.13

    C. 0.03

    D. 1.31

14. Which decimal has the **least** value?

    A. 4.8

    B. 4.98

    C. 4.89

    D. 5.9

CCSS: 4.NF.7

Directions: For questions 15 through 20, use <, >, or = to compare the decimals.

15. 0.1 ____________ 0.15

16. 0.78 ____________ 0.75

17. 3.06 ____________ 3.16

18. 0.47 ____________ 0.4

19. 1.8 ____________ 1.80

20. 0.9 ____________ 0.09

Directions: For questions 21 and 22, write the decimal represented by the shaded parts of each figure. Then use <, >, or = to compare the decimals.

21.

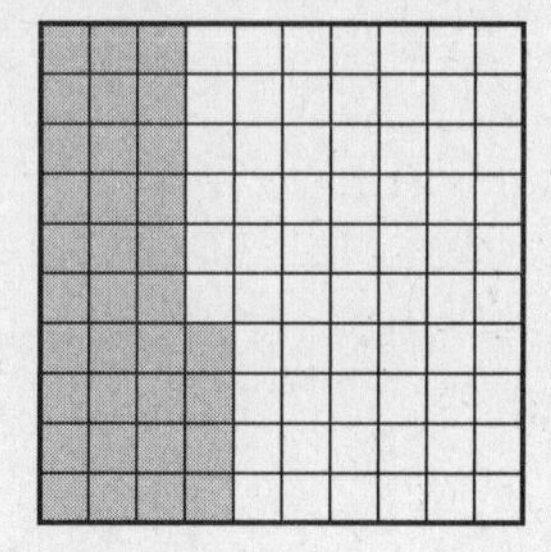

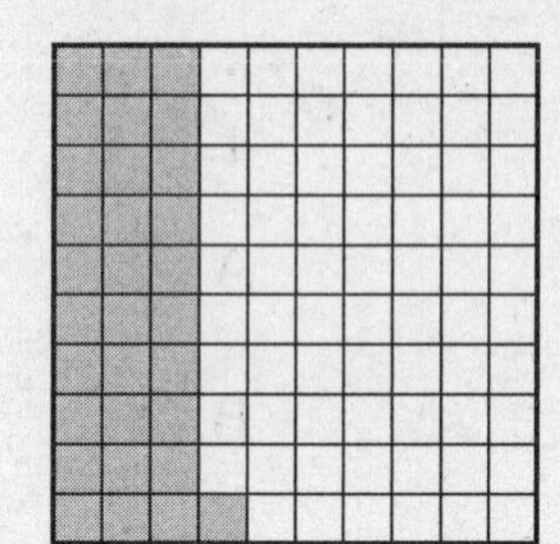

_____ _____ _____

Explain how the model shows the comparison.

____

____

22.

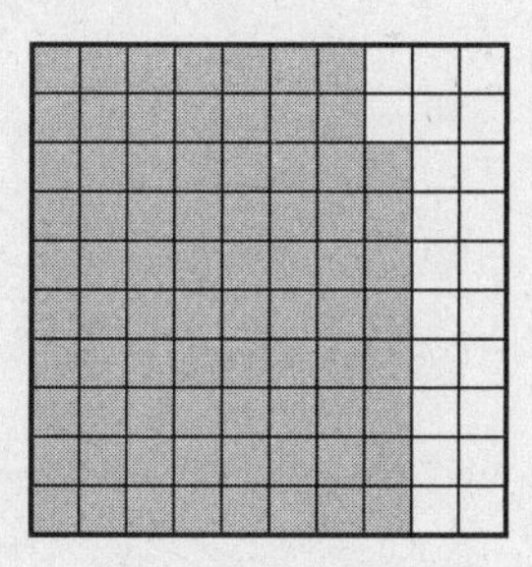

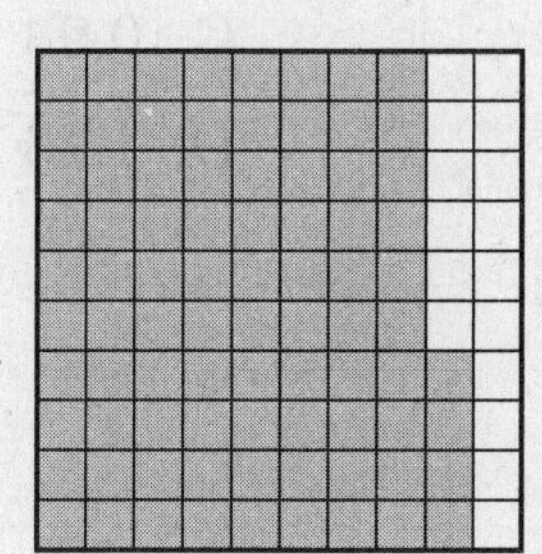

_____ _____ _____

Explain how the model shows the comparison.

____

____

Unit 3 Practice Test

Choose the correct answer.

1. Which fraction is equivalent to $\frac{1}{4}$?

   A. $\frac{2}{6}$

   B. $\frac{2}{8}$

   C. $\frac{2}{10}$

   D. $\frac{2}{12}$

2. How is 0.48 written as a fraction?

   A. $\frac{4}{10}$

   B. $\frac{8}{100}$

   C. $\frac{38}{10}$

   D. $\frac{48}{100}$

3. What is $\frac{19}{6}$ written as a mixed number?

   A. 3

   B. $3\frac{1}{6}$

   C. $3\frac{2}{6}$

   D. $6\frac{1}{3}$

4. Which is true?

   A. $\frac{5}{6} > \frac{1}{2}$

   B. $\frac{4}{10} > \frac{1}{2}$

   C. $\frac{4}{8} < \frac{1}{2}$

   D. $\frac{9}{12} < \frac{1}{2}$

5. $\frac{3}{4} + \frac{3}{4} =$

   A. $\frac{6}{8}$

   B. $1\frac{2}{8}$

   C. $1\frac{1}{4}$

   D. $1\frac{2}{4}$

6. Which decimal is less **than** 1.08?

   A. 2.08

   B. 1.88

   C. 0.98

   D. 1.18

7. Next to each model, write the fraction shown.
What equivalent fractions do the models show?

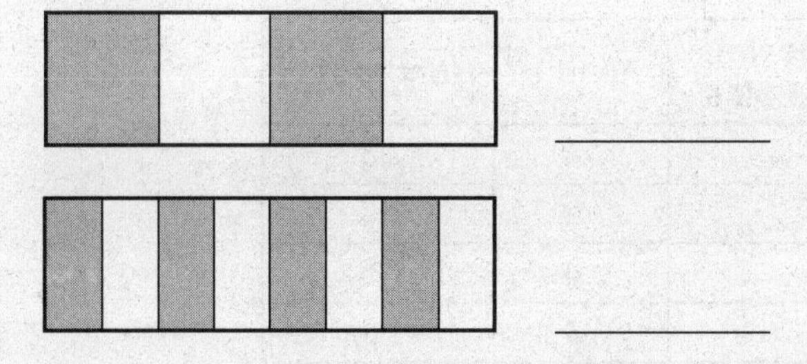

____

____

_______________ = _______________

8. The fraction $\frac{17}{8}$ is between what two whole numbers?

_______________ and _______________

To which of the two whole numbers is $\frac{17}{8}$ closer?

____

9. A package of cheese has a mass of 0.75 kilogram.

How is 0.75 written as a fraction? _______________

How is 0.75 written in words? _______________

10. $4 \times \frac{3}{10} =$ _______________

Use the decimals in the place-value table to answer questions 11 through 14.

Hundreds	Tens	Ones	Decimal Point	Tenths	Hundredths
		0	.	4	3
		1	.	2	
		0	.	4	5
		0	.	4	

11. 0.43 > ______________

12. 0.45 < ______________

13. Which of the decimals is **greatest**? ______________

14. Which of the decimals is **least**? ______________

15. Draw a picture to show that $\frac{2}{5}$ is equivalent to $\frac{4}{10}$.

Explain how you know that the fractions are equivalent.

____

____

____

16. What is the sum of $\frac{8}{10}$ and $\frac{2}{100}$?

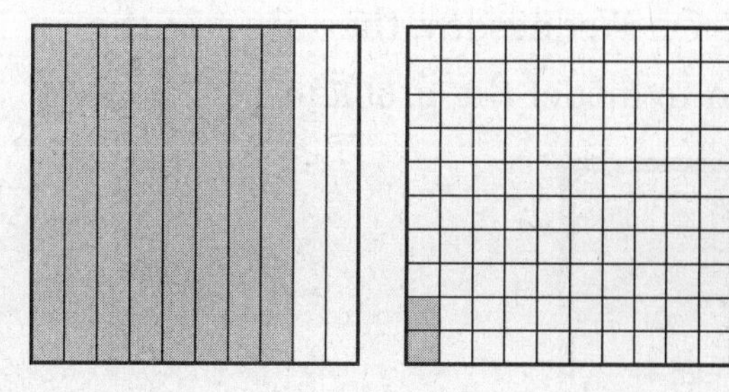

____

Explain how you got your answer.

____

____

____

17. Rewrite $\frac{3}{100}$ as the sum of unit fractions.

____

18. $3\frac{1}{8} - 1\frac{5}{8} =$ ______________

Use addition to check your answer. Explain what you did.

____

____

____

19. Draw a picture to show that $\frac{1}{2} > \frac{1}{3}$.

Explain how your drawing shows the comparison.

____

____

____

20. Isabel made a pan of brownies. On Tuesday, her family ate $\frac{3}{6}$ of the brownies. On Wednesday, they ate $\frac{2}{6}$ of the brownies. Use the pan to model the problem.

What fraction of the pan of brownies is left? ____________

21. Study the grid below.

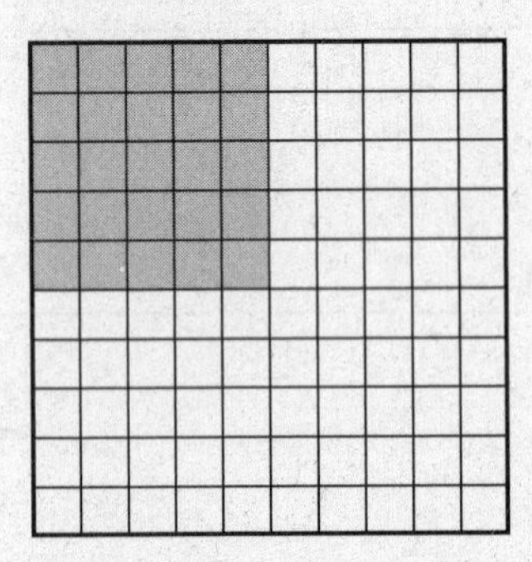

What fraction represents the shaded part? ____________

What decimal represents the shaded part? ____________

Solve each problem.

22. Benito rode $\frac{2}{8}$ of a mile to his friend's house, $\frac{1}{8}$ of a mile to the library, and $\frac{3}{8}$ of a mile back home. What is the total distance that Benito rode?

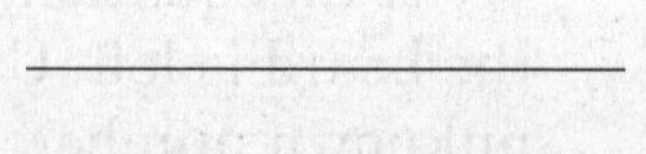

23. Mia made a large cheese sandwich to share with her brother. She put $\frac{3}{10}$ of a pound of American cheese and $\frac{2}{10}$ of a pound of Swiss cheese on the sandwich. How much cheese did Mia put on the sandwich in all?

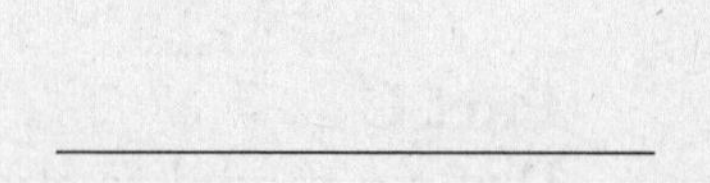

24. Jenna makes bracelets out of elastic string and beads. She uses $\frac{7}{8}$ of a foot of string for each bracelet. How much string does Jenna use to make 6 bracelets? Between what two whole numbers does your answer lie? ______________

    Check your answer by drawing a model and using repeated addition.

25. From a piece of board that is $\frac{4}{5}$ of a foot long, a carpenter sawed off a piece that was $\frac{3}{5}$ of a foot long.

Part A
Write an equation that you can use to find how much of the board is left. Use a variable such as n to represent the unknown number.

____

Part B
How much of the board is left?

____

Part C
Draw a model to check your answer.

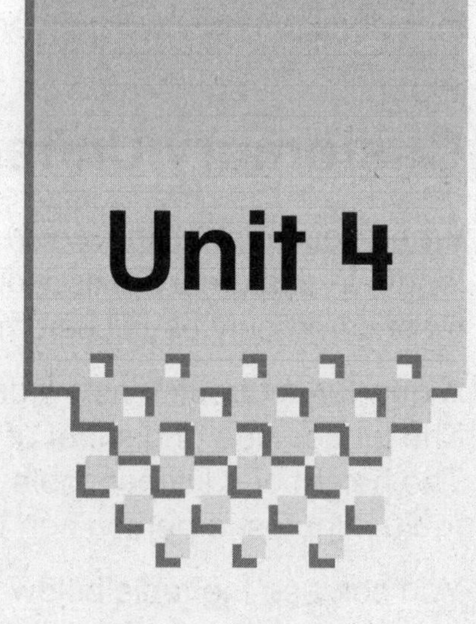

Measurement and Data

Have you ever been in a swimming pool and wondered how deep the water was, how long you were in the pool, or how much water the pool could hold? These are just a few examples of how measurement can be used in everyday life. You may not realize it, but every time you look at a clock for the time or stand on a scale to weigh yourself, you are measuring. By organizing data, you can find out things such as how many of your classmates like to go hiking, who read the most books over summer vacation, or what kinds of vegetables your classmates like to eat.

In this unit you will review length, weight, mass, capacity, and time. You will convert measurements from one unit to another within the same measurement system. You will also find the perimeter and area of rectangles and learn how to represent and interpret data by answering questions about line plots.

In This Unit

Lesson 27: Equivalent Customary Units

In this lesson, you will convert customary units of length, capacity, and weight.

Customary Units of Length

When you want to know how long an object is, you measure its **length**. To measure length in customary units, you can use a **ruler**, a **yardstick**, or a **tape measure**. The most commonly used customary units of length are **inch**, **foot**, **yard**, and **mile**.

A quarter is about 1 inch wide.
The longer side of a sheet of notebook paper is about 1 foot.
The height of a kitchen table is about 1 yard.
In 20 minutes, a person can walk about 1 mile.

You can use the table below to help you convert from one customary unit of length to an equivalent number of smaller customary units of length.

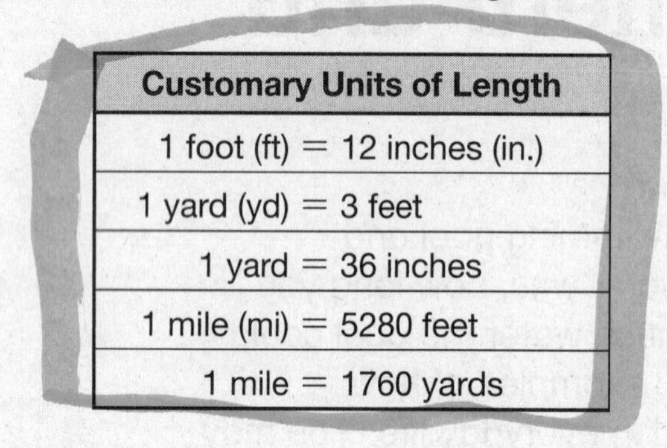

Customary Units of Length
1 foot (ft) = 12 inches (in.)
1 yard (yd) = 3 feet
1 yard = 36 inches
1 mile (mi) = 5280 feet
1 mile = 1760 yards

To convert from a larger unit to an equivalent number of smaller units, you multiply because there will be more of the smaller units.

Examples

36 ft = ____?____ in.
1 ft = 12 in. and 36 × 12 in. = 432 in.
So 36 ft = 432 in.

6 yd = ____?____ ft
1 yd = 3 ft and 6 × 3 ft = 18 ft
So 6 yd = 18 ft.

5 yd = ____?____ in.
1 yd = 36 in. and 5 × 36 in. = 180 in.
So 5 yd = 180 in.

3 mi = ____?____ yd
1 mi = 1760 yd and 3 × 1760 yd = 5280 yd
So 3 mi = 5280 yd.

CCSS: 4.MD.1

Customary Units of Capacity

When you want to know how much liquid an object can hold, you measure its **capacity**. To measure capacity in customary units, you can use **measuring cups**. The most commonly used customary units of capacity are **cup**, **pint**, **quart**, and **gallon**.

A coffee mug holds about 1 cup.
A soup bowl holds about 1 pint.
A serving bowl holds about 1 quart.
A large container of milk holds about 1 gallon.

You can use the table below to help you convert from one customary unit of capacity to an equivalent number of smaller customary units of capacity.

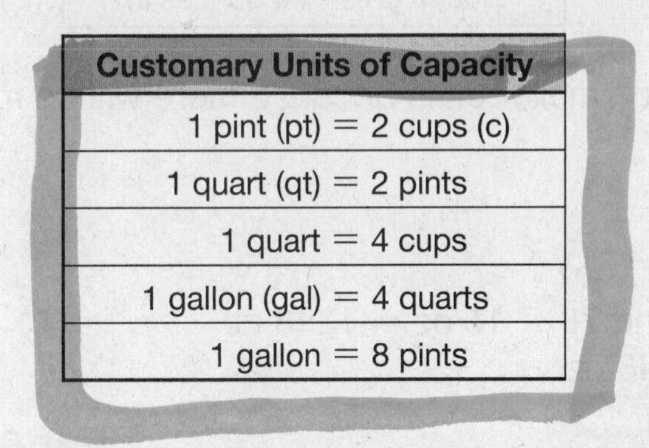

Customary Units of Capacity
1 pint (pt) = 2 cups (c)
1 quart (qt) = 2 pints
1 quart = 4 cups
1 gallon (gal) = 4 quarts
1 gallon = 8 pints

You multiply to convert to smaller units because there will be more smaller units.

Examples

15 pt = ___?___ c
1 pt = 2 c and 15 × 2 c = 30 c
So 15 pt = 30 c.

32 qt = ___?___ pt
1 qt = 2 pt and 32 × 2 p = 64 pt
So 32 qt = 64 pt.

17 qt = ___?___ c
1 qt = 4 c and 17 × 4 c = 68 c
So 17 qt = 68 c.

59 gal = ___?___ qt
1 gal = 4 qt and 59 × 4 qt = 236 qt
So 59 gal = 236 qt.

25 gal = ___?___ pt
1 gal = 8 pt and 25 × 8 pt = 200 pt
So 25 gal = 200 pt.

Customary Units of Weight

When you want to know how heavy an object is, you measure its **weight**. To measure weight in customary units, you can use a **scale**. The most commonly used customary units of weight are **ounce** and **pound**.

A slice of bread weighs about 1 ounce.
A loaf of bread weighs about 1 pound.

You can use the table below to help you convert from one customary unit of weight to an equivalent number of smaller customary units of weight.

Customary Units of Weight
1 pound (lb) = 16 ounces (oz)

You multiply to convert to smaller units because there will be more smaller units.

Example

78 lb = ____?____ oz
1 lb = 16 oz and 78 × 16 oz = 1248 oz
So 78 lb = 1248 oz.

Practice

Directions: For questions 1 through 6, use the tables of equivalents in the lesson to complete the conversion tables.

1.

Yards	Inches
1	36
2	72
3	108
4	144
5	180
6	216

2.

Gallons	Quarts
1	4
2	8
3	12
4	16
5	20
6	24

CCSS: 4.MD.1

3.

Pounds	Ounces
1	
2	
3	
4	
5	
6	
7	
8	
9	
10	

4.

Pints	Cups
1	
2	
3	
4	
5	
6	
7	
8	
9	
10	

5.

Feet	Inches
1	
2	
3	
4	
5	
6	
7	
8	
9	
10	

6.

Quarts	Cups
1	
2	
3	
4	
5	
6	
7	
8	
9	
10	

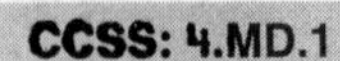

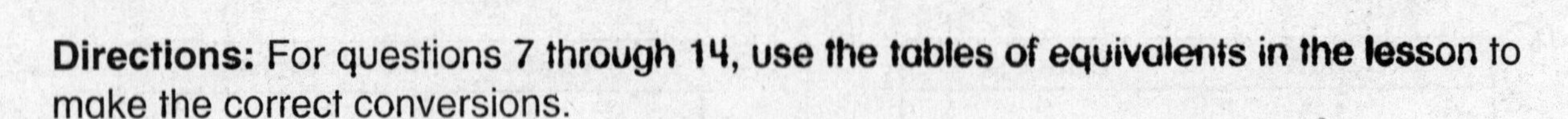

Directions: For questions 7 through 14, use the tables of equivalents in the lesson to make the correct conversions.

7. 36 ft = ______________ in.

8. 8 lb = ______________ oz

9. 12 pt = ______________ c

10. 24 yd = ______________ ft

11. 30 lb = ______________ oz

12. 19 qt = ______________ pt

13. 7 yd = ______________ in.

14. 11 gal = ______________ qt

15. 4 ft =

A. 12 in.
B. 24 in.
C. 48 in.
D. 48 yd

16. 10 lb =

A. 16 oz
B. 160 oz
C. 160 in.
D. 160 yd

17. Ron sold a bushel of onions that weighed 64 pounds. How many ounces did the bushel of onions weigh?

A. 4
B. 16
C. 1024
D. 64,000

18. 13 pt =

A. 25 c
B. 26 c
C. 26 pt
D. 26 gal

19. 22 gal =

A. 44 qt
B. 88 c
C. 88 pt
D. 88 qt

20. Which weighs the most?

A. a bag weighing 2 lb 8 oz
B. a 3-pound bag from which 12 oz has been removed
C. a bag containing 2 lb of green apples and 12 oz of red apples
D. a bag weighing 42 oz

CCSS: 4.MD.1

21. Would it take a **greater** number of cups or pints to measure the capacity of a bathtub? Explain your answer.

____

____

____

22. Would it take a **greater** number of yards or miles to measure the distance from Lakeview to Sun City? Explain your answer.

____

____

____

23. Would it take a **greater** number of ounces or pounds to measure the weight of an elephant? Explain your answer.

____

____

____

24. Would it take a **greater** number of quarts or gallons to measure the capacity of a swimming pool? Explain your answer.

____

____

____

Lesson 28: Equivalent Metric Units

In this lesson, you will convert metric units of length, capacity, and mass.

Metric Units of Length

To measure length in metric units, you can use a **ruler**, a **meterstick**, or a **tape measure**. The most commonly used metric units of length are **centimeter**, **meter**, and **kilometer**.

The width of a large paper clip is about 1 centimeter.
The length of a adult's baseball bat is about 1 meter.
In 15 minutes, most people can walk about 1 kilometer.

You can use the table below to help you convert from one metric unit of length to an equivalent number of smaller metric units of length.

Metric Units of Length
1 meter (m) = 100 centimeters (cm)
1 kilometer (km) = 1000 meters

To convert from a larger unit to an equivalent number of smaller units, you multiply because there will be more of the smaller units.

Examples

9 m = ____?____ cm
1 m = 100 cm and 9 × 100 cm = 900 cm
So 9 m = 900 cm.

6 km = ____?____ m
1 km = 1000 m and 6 × 1000 m = 6000 m
So 6 km = 6000 m.

There are 100 centimeters in a meter, so each centimeter is $\frac{1}{100}$ of a meter. You can use a decimal to express some number of centimeters as a fraction of a meter.

Example

Use a decimal to write 78 centimeters as a number of meters.

78 centimeters = $\frac{78}{100}$ meter

$\frac{78}{100} = 0.78$

So 78 centimeters = 0.78 meter.

CCSS: 4.NF.6, 4.MD.1

Metric Units of Capacity

To measure capacity in metric units, you can use **measuring cups**.
The most commonly used metric units of capacity are **liter** and **milliliter**.

The capacity of an eyedropper is about 2 milliliters.
The capacity of a large bottle of cola is about 1 liter.

You can use the table below to help you convert from one metric unit of capacity to an equivalent number of smaller metric units of capacity.

Metric Units of Capacity
1 liter (L) = 1000 milliliters (mL)

You multiply to convert to smaller units because there will be more smaller units.

Example

8 L = ___?___ mL
1 L = 1000 mL and 8 × 1000 mL = 8000 mL
So 8 L = 8000 mL.

Units of Metric Mass

You can use metric units of **mass** to measure how much matter an object has.
To measure mass, you use a **scale**. The most commonly used metric units of mass are **gram** and **kilogram**.

A paper clip has a mass of about 1 gram.
A pineapple has a mass of about 1 kilogram.

You can use the table below to help you convert from one metric unit of mass to an equivalent number of smaller metric units of mass.

Metric Units of Mass
1 kilogram (kg) = 1000 grams (g)

You multiply to convert to smaller units because there will be more smaller units.

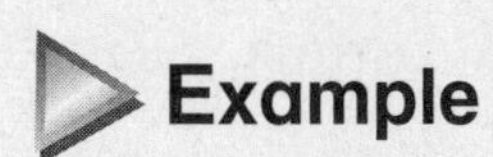

Example

5 kg = ___?___ g
1 kg = 1000 g and 5 × 1000 g = 5000 g
So 5 kg = 5000 g.

Practice

Directions: For questions 1 through 4, use the tables of equivalents in the lesson to complete the conversion tables.

1.

Meter	Centimeters
1	
2	
3	
4	
5	
6	
7	
8	
9	
10	

2.

Liters	Milliliters
1	
2	
3	
4	
5	
6	
7	
8	
9	
10	

3.

Kilograms	Grams
1	
2	
3	
4	
5	
6	
7	
8	
9	
10	

4.

Kilometers	Meters
1	
2	
3	
4	
5	
6	
7	
8	
9	
10	

CCSS: 4.NF.6, 4.MD.1

Directions: For questions 5 through 12, use the tables of equivalents in the lesson to make the correct conversions.

5. 8 m = ______________ cm

6. 2 kg = ______________ g

7. 5 km = ______________ m

8. 9 L = ______________ mL

9. 4 kg = ______________ g

10. 3 L = ______________ mL

11. 7 m = ______________ cm

12. 4 km = ______________ m

13. 3 m =

    A. 30 cm

    B. 300 cm

    C. 3000 cm

    D. 3000 km

14. 8 kg =

    A. 80 g

    B. 800 g

    C. 8000 g

    D. 8000 m

15. Julie's dog has a mass of 3 kilograms. What is the mass of Julie's dog in grams?

    A. 30 grams

    B. 300 grams

    C. 3000 grams

    D. 30,000 grams

16. 2 L =

    A. 2000 mL

    B. 2000 cm

    C. 2000 g

    D. 2000 m

17. 8 km =

    A. 8000 cm

    B. 8000 m

    C. 8000 g

    D. 8000 kg

18. Which has the greatest mass?

    A. a bag with a mass of 3 kg

    B. a 4-kg bag from which 2500 g has been removed

    C. a bag containing a 2-kg chicken and a 2500-g package of cheese

    D. a bag weighing 3500 g

Directions: For questions 19 through 21, use a decimal to write the given number of centimeters as a number of meters.

19. 23 cm = __________ 20. 5 cm = __________ 21. 142 cm = __________

22. Would it take a **greater** number of milliliters or liters to measure the capacity of a pond? Explain your answer.

____

____

____

23. Would it take a **greater** number of meters or centimeters to measure the length of your desk? Explain your answer.

____

____

____

24. Would it take a **greater** number of kilograms or grams to measure the mass of a puppy? Explain your answer.

____

____

____

25. Would it take a **greater** number of kilometers or meters to measure the length of a river? Explain your answer.

____

____

____

CCSS: 4.MD.1

Lesson 29: Equivalent Units of Time

Time tells you what part of the day or night it is or how long it takes for an event to occur. To measure time, you can use an **analog clock** or a **digital clock**. The most commonly used units of time are **second**, **minute**, and **hour**.

It takes about 1 second to blink your eyes.
It takes about 1 minute to walk 1 block.
It takes about 1 hour to drive 50 miles.

You can use the table below to help you convert from one unit of time to an equivalent number of smaller units of time.

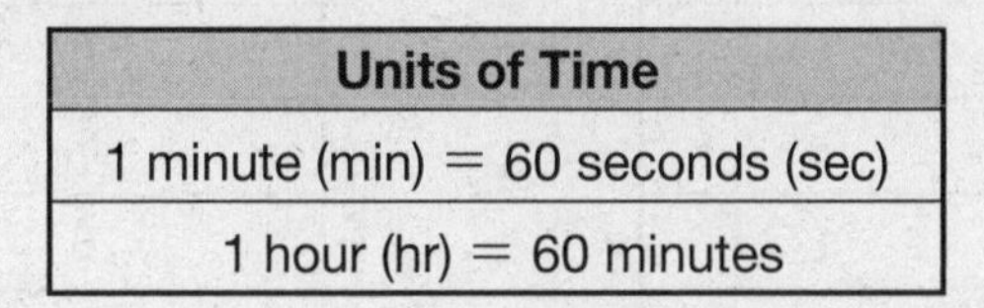

Units of Time
1 minute (min) = 60 seconds (sec)
1 hour (hr) = 60 minutes

Examples

20 min = ___?___ sec
1 min = 60 sec and 20 × 60 sec = 1200 sec
So 20 min = 1200 sec.

15 hr = ___?___ min
1 hr = 60 min and 15 × 60 min = 900 min
So 15 hr = 900 min.

You can use the table above and an extra step to convert a number of hours to an equivalent number of seconds.

Example

3 hr = ___?___ sec
1 hr = 60 min and 1 min = 60 sec
60 min = 60 × 60 sec = 3600 sec
So 1 hr = 3600 sec.

3 hr = 3 × 3600 sec = 10,800 sec
So 3 hr = 10,800 sec.

Practice

Directions: For questions 1 and 2, use the table of equivalents in the lesson to complete the conversion tables.

1.

Hours	Minutes
1	
2	
3	
4	
5	
6	
7	
8	
9	
10	
11	
12	
13	
14	
15	
16	
17	
18	
19	
20	

2.

Minutes	Seconds
1	
2	
3	
4	
5	
6	
7	
8	
9	
10	
11	
12	
13	
14	
15	
16	
17	
18	
19	
20	

CCSS: 4.MD.1

Directions: For questions 3 through 10, use the table of equivalents in the lesson to make the correct conversions.

3. 12 min = ______________ sec

4. 4 hr = ______________ min

5. 52 min = ______________ sec

6. 31 hr = ______________ min

7. 7 min = ______________ sec

8. 60 hr = ______________ min

9. 48 min = ______________ sec

10. 24 hr = ______________ min

11. How many seconds are in 70 minutes?

    A. 42
    B. 420
    C. 4200
    D. 42,000

12. How many minutes are in 13 hours?

    A. 78
    B. 780
    C. 7800
    D. 78,000

13. Ben goes to a movie that is $2\frac{1}{2}$ hours long. How many minutes are equal to $2\frac{1}{2}$ hours?

    A. 60 minutes
    B. 90 minutes
    C. 120 minutes
    D. 150 minutes

14. How many minutes are in 6 hours?

    A. 36
    B. 360
    C. 3600
    D. 36,000

15. How many seconds are in 40 minutes?

    A. 24
    B. 240
    C. 2400
    D. 24,000

16. Which amount of time is greatest?

    A. 2 hours and 30 minutes
    B. 1 hour and 75 minutes
    C. 2 hours and 120 seconds
    D. 2 hours, 15 minutes, and 120 seconds

Directions: For questions 17 and 18, write the missing units of time. Use *seconds*, *minutes*, or *hours*.

17. A fourth grader can walk 4 kilometers in about 1 ________, or 60 ________.

    Explain your answer.

    ____

    ____

    ____

18. It takes about 2 ________, or 120 ________, to brush your teeth.

    Explain your answer.

    ____

    ____

    ____

19. Would it take a **greater** number of seconds or minutes to represent the amount of time Rasheed spent doing homework?

    Explain your answer.

    ____

    ____

    ____

20. Would it take a **greater** number of minutes or hours to represent the amount of time Mia spent playing soccer?

    Explain your answer.

    ____

    ____

    ____

Lesson 30: Solving Measurement Problems

When solving measurement problems, you may find it helpful to **make a table of** equivalents for the units given. For some problems, you can draw a number line to show equal parts of units. Sometimes solving a problem involves more than one step, so be sure to answer the question that the problem asks.

Example

Evan spent 3 hours at the art museum. How many minutes did he spend at the museum?

3 hr = ___?___ min
1 hr = 60 min
3 × 60 min = 180 min

So Evan spent 180 min at the art museum.

Example

Roseanne needs 1.5 lb of sliced cheese to make sandwiches for lunch. She bought a package of cheese slices marked 1.28 lb. Does this package have enough cheese for her to use in making lunch?

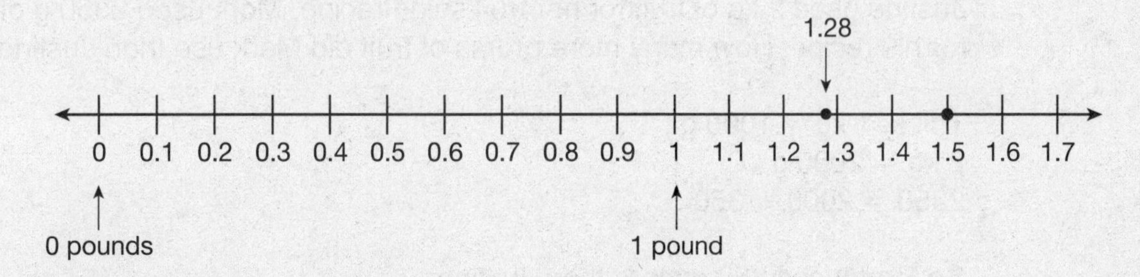

You can use the number line to locate the point for 1.28.
Is 1.28 greater or less than 1.5?

The point for 1.28 is to the left of 1.5, between 1.2 and 1.3.
Therefore,1.28 is less than 1.5.

So there is not enough cheese for Roseanne to use in making lunch.

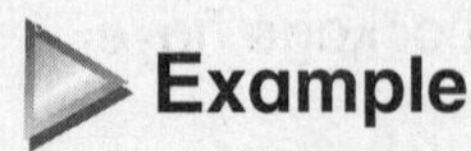

Example

Lily's hike was 2 miles long. She stopped to rest after hiking for $\frac{3}{4}$ mi. How much farther did she have to go after stopping to rest?

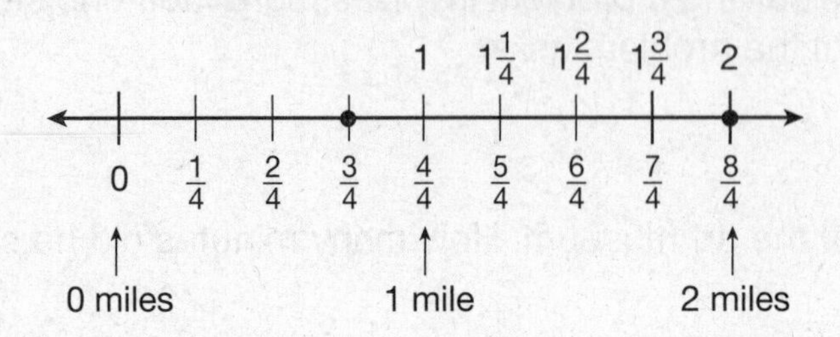

You can use the number line to count that there are $\frac{5}{4}$ left to go.

$\frac{5}{4} = \frac{4}{4} + \frac{1}{4}$, or $1\frac{1}{4}$ miles.

So there are $1\frac{1}{4}$ mi left to go.

You can also subtract:

$2 = \frac{8}{4}$ and $\frac{8}{4} - \frac{3}{4} = \frac{5}{4}$.

Example

Justine used 2 kg of fruit for her fruit salad recipe. Mark used 2350 g of fruit for his recipe. How many more grams of fruit did Mark use than Justine?

Think: 1 kg = 1000 g
2 kg = 2000 g
2350 = 2000 + 350

So Mark used 350 g more than Justine.

Example

A chef had a 3-liter bottle of olive oil. He used the same amount of olive oil each day. If the oil was used up in 5 days, how many milliliters of olive oil did the chef use each day?

Think: 1 L = 1000 mL

3 L = 3000 mL

$3000 \div 5 = 600$

So the chef used 600 mL of olive oil each day.

Example

Ricky bought 3 shirts. The first shirt cost $15. The second shirt cost $3 more than the first shirt. The third shirt cost $1 less than the second shirt. How much did the third shirt cost?

$15 + $3 = cost of second shirt

$15 + $3 = ☐

$15 + $3 = $18

The second shirt cost $18.

$18 − $1 = cost of third shirt

$18 − $1 = ☐

$18 − $1 = 17

So the third shirt cost $17.

Example

Sonja had a piece of fabric that was 7 feet long. Stuart had a piece of fabric that was 5 feet 10 inches long. Robert's piece of fabric was 2 yards 8 inches long. Who had the longest piece of fabric?

Convert the larger units to inches in order to compare like units.

Sonja's fabric: 1 foot = 12 inches and 7 feet = 7 × 12 inches = 84 inches

Stuart's fabric: 5 feet = 5 × 12 inches = 60 inches

60 inches + 10 inches = 70 inches

Robert's fabric: 1 yard = 36 inches   2 yards = 2 × 36 inches = 72 inches

72 inches + 8 inches = 80 inches

So Robert had the longest piece of fabric.

Practice

Solve each problem.

1. Maria's new puppy weighs 3 pounds. How many ounces does Maria's new puppy weigh?

   ____

2. Brandon completed a 1-kilometer race in 8 minutes. He ran the same distance each minute. How many meters did Brandon run each minute?

   ____

3. On April 1, Tanesha planted one daffodil bulb and one tulip bulb. On June 1 the daffodil was $\frac{7}{12}$ of a foot tall and the tulip was $\frac{5}{12}$ of a foot tall. How much taller was the daffodil than the tulip on June 1?

   ____

4. Carlos spent 2 hours at the gym. He spent 40 minutes on the treadmill and 30 minutes working with weights. He spent the rest of the time swimming. How many minutes did Carlos spend swimming?

   ____

   Explain how you got your answer.

   ____

   ____

   ____

5. Sue caught a fish that weighed 8 pounds. Her brother caught a fish that weighed $\frac{3}{4}$ as much. How much did her brother's fish weigh?

   ____

6. Mr. Chan drove a distance of 786 miles in 3 days. He drove the same number of miles each day. How many miles did Mr. Chan drive each day?

   ____

7. Walter bought $\frac{5}{8}$ pound of American cheese and $\frac{2}{8}$ pound of Swiss cheese. How much cheese did Walter buy?

8. In science class, Arnold poured 97 mL of red-tinted water into a 1-L jar. Then he filled the jar with blue-tinted water. How much blue-tinted water did Arnold pour into the jar?

   ____

   Explain how you got your answer.

   ____

   ____

   ____

9. Each person in Denzel's family drinks $\frac{2}{3}$ of a pint of milk each day. There are 4 people in Denzel's family. How many pints of milk does Denzel need to buy for his family for Monday?

____

10. Mercedes bought a 2-kilogram package of ground turkey. She made a 290-gram turkey burger for dinner. How many grams of turkey are left in the package?

____

11. Paul bought a CD player that cost $82. The tax on the player was $6. Paul paid with a $100 bill. How much change did he get?

____

12. Betty has two hair ribbons. The red ribbon is 40.1 cm long, and the yellow ribbon is 40.08 cm long. Which color ribbon is shorter?

____

Explain how you got your answer.

____

____

____

CCSS: 4.MD.3

Lesson 31: Perimeter

The distance around a two-dimensional figure is its **perimeter (*P*)**. Perimeter can be measured in customary or metric units. To find the perimeter of a figure, add the lengths of all its sides.

***P* = side + side + side + side...(add as many sides as the figure has)**

You can also use the formula $P = (2 \times l) + (2 \times w)$ to find the perimeter of a rectangle, where l is the length and w is the width.

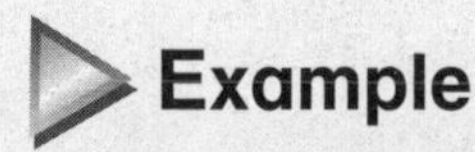

Example

What is the perimeter (P) of this rectangle?

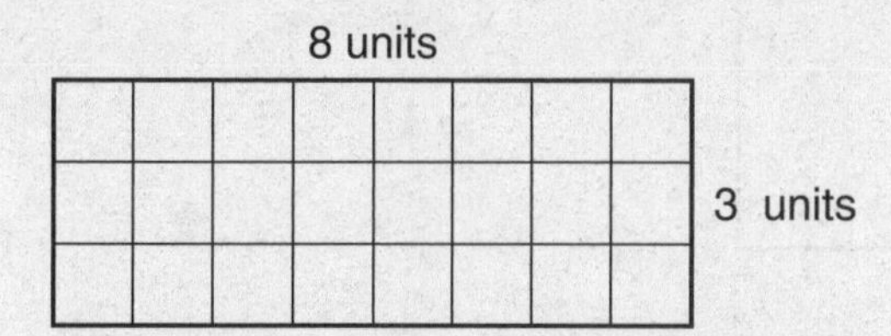

$$\begin{aligned} P &= \text{side} + \text{side} + \text{side} + \text{side} \\ &= 8 + 3 + 8 + 3 \\ &= 11 + 11 \\ &= 22 \end{aligned}$$

$$\begin{aligned} P &= (2 \times l) + (2 \times w) \\ &= (2 \times 8) + (2 \times 3) \\ &= 16 + 6 \\ &= 22 \end{aligned}$$

The perimeter is 22 units.

In a square, all of the sides have the same length. Because of this, you can use the formula $P = 4 \times s$ to find the perimeter of a square, where s is a side of the square.

Example

What is the perimeter (P) of this square?

$$\begin{aligned} P &= 4 \times s \\ &= 4 \times 3 \\ &= 12 \end{aligned}$$

The perimeter is 12 units.

Sometimes you will have to measure to find the dimensions of a figure before you can find its perimeter. If the unit of measurement is given to you, make sure you measure using that unit. If no unit is given, choose a unit that is suitable.

Example

Measure the dimensions of the following rectangle. Then use those dimensions to find its perimeter.

Start by choosing a suitable unit.
Inches and centimeters are both suitable units. Use centimeters.

Measure the length and width of the rectangle.
length (l) = 6 cm
width (w) = 4 cm

Use a formula to find the perimeter of the rectangle.

$$\boldsymbol{P = (2 \times l) + (2 \times w)}$$
$$= (2 \times 6) + (2 \times 4)$$
$$= 12 + 8$$
$$= 20$$

The perimeter of the rectangle is 20 centimeters.

CCSS: 4.MD.3

You can use the perimeter of a square to find the lengths of the sides.

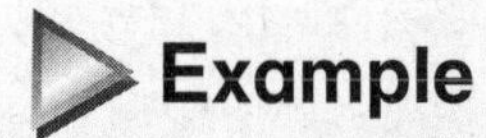

Example

The perimeter of a square is 24 inches. What is the length of each side?

Use the formula $\boldsymbol{P = 4 \times s}$.
Think: Each side is the same number of inches.

$4 \times \square = 24$

$4 \times 6 = 24$

So each side is 6 inches long.

You can also think: Each side is $\frac{1}{4}$ of the perimeter, and $\frac{1}{4}$ means to divide the whole by 4.

$24 \div 4 = 6$

You can use the perimeter and one of the dimensions of a rectangle to find the other dimension.

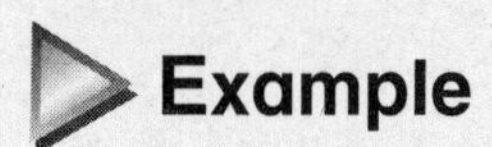

Example

The perimeter of a rectangle is 48 centimeters. The length of the rectangle is 14 centimeters. What is the width of the rectangle?

Use the formula for the perimeter of a rectangle and fill in the numbers you know.

$\boldsymbol{P = (2 \times l) + (2 \times w)}$
$48 = (2 \times 14) + (2 \times w)$
$48 = 28 + 2w$

Think: $28 +$ what number $= 48$?
$28 + 20 = 48$
and $20 = 2 \times w$, so $w = 10$.

The width of the rectangle is 10 centimeters.

You can sketch a model to help you solve a perimeter problem when no model is given.

Example

A rectangular rug is 10 feet long and 6 feet wide. What is the perimeter of the rug?

Sketch a rectangle. Label the length “10 ft” and label the width “6 ft.”

$\mathbf{P = (2 \times \mathit{l}) + (2 \times \mathit{w})}$

$= (2 \times 10) + (2 \times 6)$

$= 20 + 12$

$= 32$

The perimeter of the rug is 32 feet.

Practice

Directions: For questions 1 through 4, find the perimeter of each figure. Measure the lengths of the sides in centimeters.

1.

$P =$ ____________

2. 8 units

10 units

$P =$ 36 units

3. a square that is 8 yards on each side

$P =$ 32 Label

4. a rectangle that is 18 cm long and 11 cm wide

$P =$ 58

5. Which figure has a perimeter of 44 feet?

A. 10 ft, 8 ft

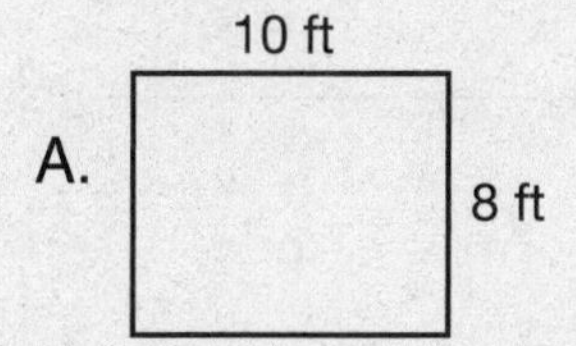

C. 15 ft, 9 ft

B. 11 ft, 11 ft

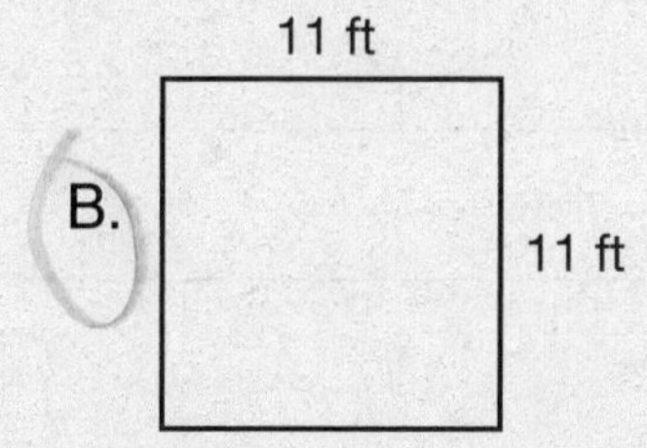

D. 8 ft, 8 ft

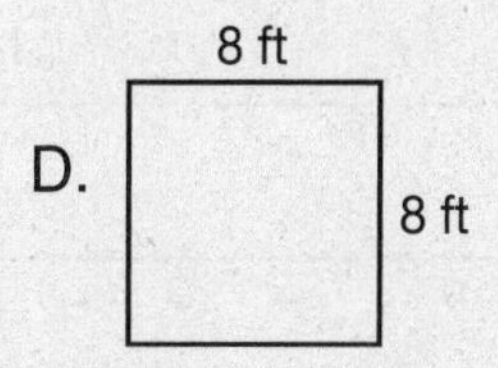

6. Which figure has a perimeter of 32 meters?

A. 6 m, 6 m

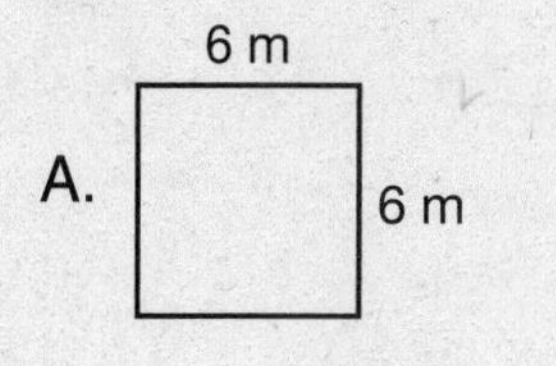

C. 10 m, 10 m

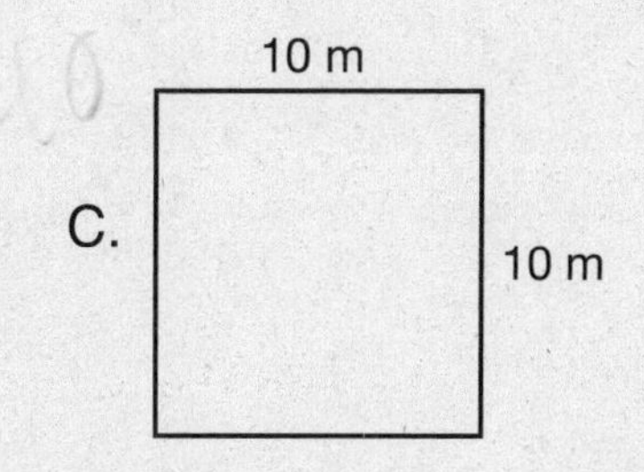

B. 12 m, 5 m

D. 13 m, 3 m

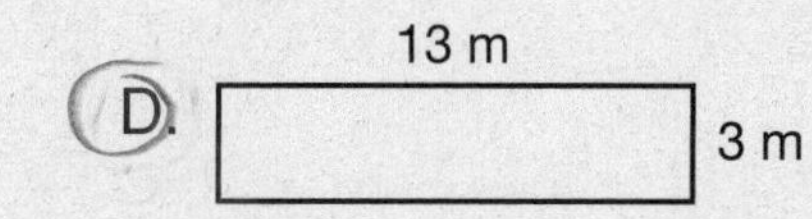

7. Leona is putting a border around a quilt. The longer side of the quilt measures 11 feet. The shorter side measures 8 feet. What is the length of the border that Leona is putting around her quilt?

____

8. Arnie is helping his dad put a fence around their rectangular garden patch. The longer side of the patch measures 8 meters. The shorter side measures 5 meters. What is the length of the fence that Arnie and his dad need?

____

9. The perimeter of a square stick-on note is 12 inches. What is the length of each side?

____

10. The perimeter of a rectangular sheet of paper is 100 centimeters. The width of the paper is 20 centimeters. What is the length of the paper?

____

Explain how you got your answer.

____

____

____

CCSS: 4.MD.3

Lesson 32: Area

Area (***A***) is the size of a region in terms of the number of unit squares that can completely cover the region. The length of each side of a unit square is 1 unit, and the area of that square is 1 square unit. To find the area of a rectangle, you can count the number of square units that cover the rectangle without any overlapping or gaps. You can also use the formula for the area of a rectangle:

$A = l \times w$ or the area of square: $A = s \times s$

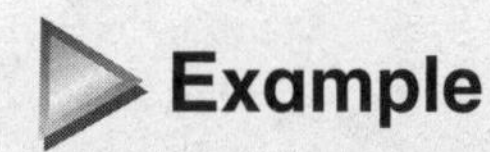

Example

What is the area (A) of this rectangle?

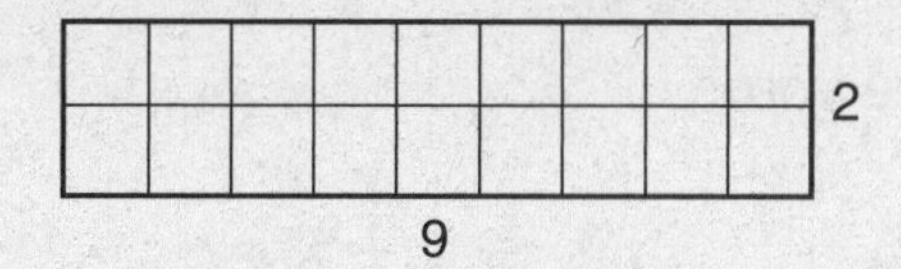

$A = l \times w$
$= 9 \times 2$
$= 18$

The area of the rectangle is 18 square units.
Notice that the product is the same number of square units you get when you count the unit squares.

Example

What is the area (A) of this rectangle?

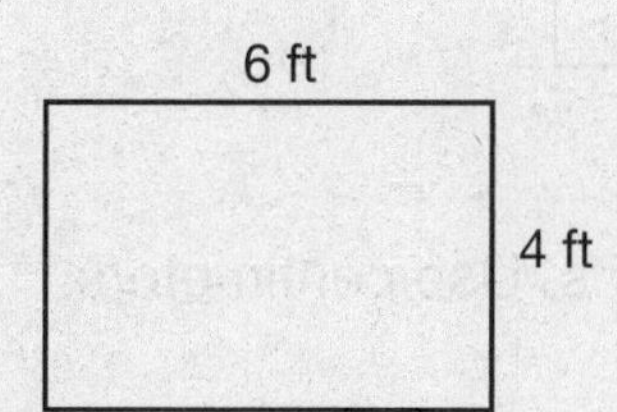

$A = l \times w$
$= 6 \times 4$
$= 24$

The area of the rectangle is 24 square feet.

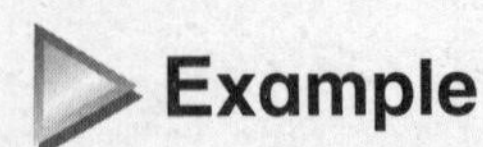

What is the area (A) of this square?

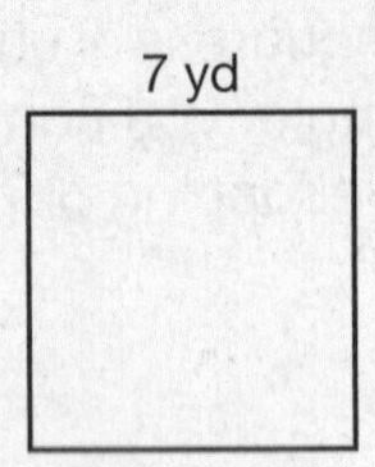

$$\begin{aligned} \boldsymbol{A} &= \boldsymbol{s} \times \boldsymbol{s} \\ &= 7 \times 7 \\ &= 49 \end{aligned}$$

The area of the square is 49 square yards.

Sometimes you will have to measure to find the dimensions of a figure before you can find its area.

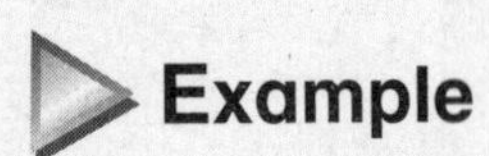

Measure the dimensions of the following rectangle. Then use those dimensions to find its area.

Start by choosing a suitable unit.
Both inches and centimeters are suitable units. Use centimeters.

Measure the length and width of the rectangle.
length (l) = 8 cm
width (w) = 2 cm

Use a formula to find the area of the rectangle.

$$\begin{aligned} \boldsymbol{A} &= \boldsymbol{l} \times \boldsymbol{w} \\ &= 8 \times 2 \\ &= 16 \end{aligned}$$

The area of the rectangle is 16 square centimeters.

CCSS: 4.MD.3

You can use the area and one of the dimensions of a rectangle to find the other dimension.

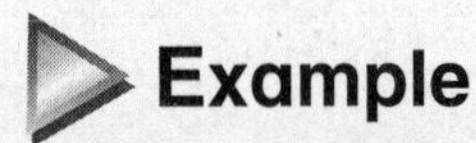

Example

The area of a rectangle is 45 square meters. The length of the rectangle is 9 meters. What is the width of the rectangle?

Use the formula for the area of a rectangle and fill in the numbers you know.

$\boldsymbol{A = l \times w}$

$45 = 9 \times w$

Think: $9 \times$ what number $= 45$?

$9 \times \quad 5 \quad = 45$

so $w = 5$.

The width of the rectangle is 5 meters.

You can use the area of a square to find the lengths of the sides.

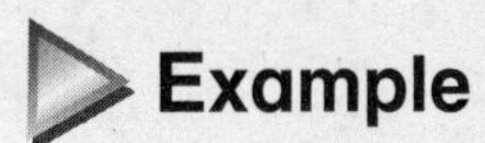

Example

The area of a square is 81 square inches. What is the length of each side?

Use the formula $\boldsymbol{A = s \times s}$.

You know that both sides are the same length and that the area is 81.

Think: What number multiplied by itself equals 81?

$81 = 9 \times 9$

So each side is 9 inches long.

You can use the area formula to solve word problems that involve the area of squares and other rectangles.

Example

A rectangular flower garden is 15 feet long and 9 feet wide. What is the area of the garden?

$$\begin{aligned} \boldsymbol{A} &= \boldsymbol{l} \times \boldsymbol{w} \\ &= 15 \times 9 \\ &= 135 \end{aligned}$$

The area of the garden is 135 square feet.

Practice

Directions: For questions 1 through 3, measure the lengths of the sides of each figure using the given unit. Then find the area.

1. centimeters

A = ______________

2. inches

A = ______________

3. centimeters

$A =$ ______________

Directions: For questions 4 through 6, find the area of each figure.

4. 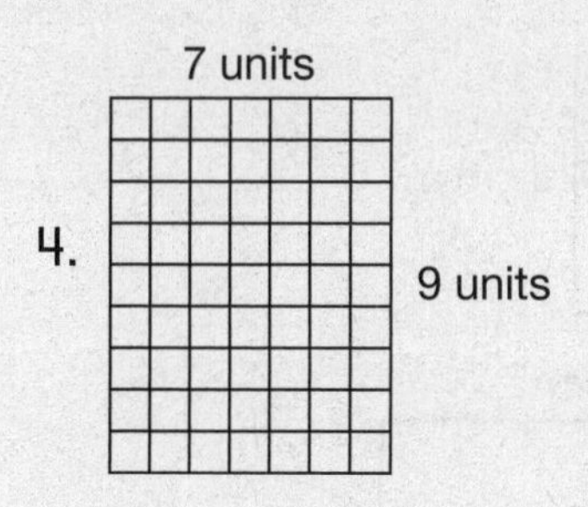

$A =$ ______________

5. a square that is 15 yards on each side

$A =$ ______________

6. a rectangle that is 13 cm long and 9 cm wide

$A =$ ______________

7. Which figure has an area of 70 square inches?

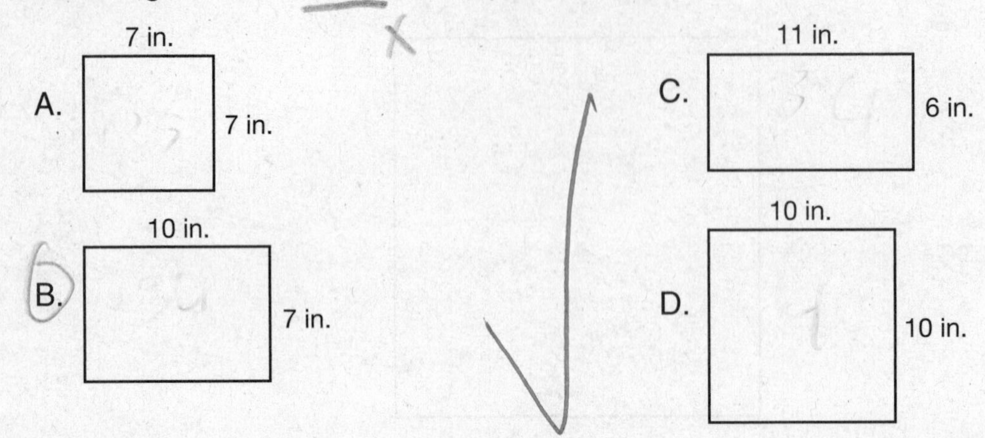

8. Which figure has an area of 126 square centimeters?

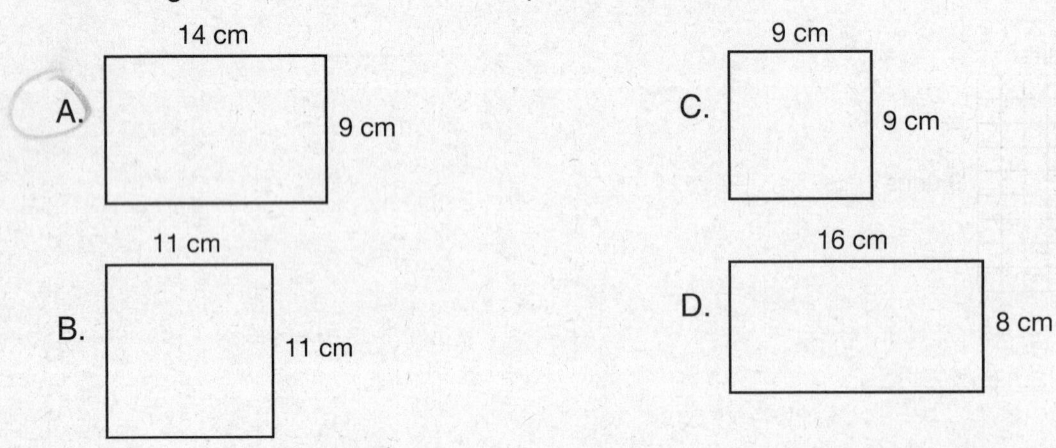

CCSS: 4.MD.3

9. Linda wants to make a blanket for the top of her bed. Her bed measures 7 feet long by 4 feet wide. How many square feet must the blanket be to cover the top of her bed?

   28?

10. Larry is going to paint his garage door. The door is in the shape of a square with each side measuring 3 meters. How many square meters will Larry need to paint?

   12?

11. The area of a rectangular photograph is 150 square centimeters. The length of the photograph is 15 centimeters. What is the width of the photograph?

   10?

12. The area of a square CD case is 100 square centimeters. What is the length of each side?

   25

Explain how you got your answer.

because 25 X 4 = 100

Lesson 33: Line Plots

A **line plot** is used to show how closely grouped together or how spread out over a range the data are. Each X mark on the line plot shows the number of times the data value occurs.

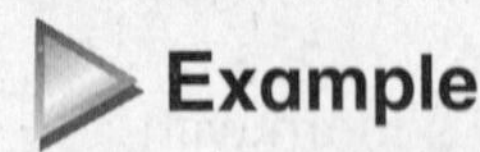

Example

Mr. Campbell asked his fourth-grade students how many pets they have.
Here are the results:
3, 2, 0, 0, 1, 1, 1, 4, 6, 0, 0, 2, 1, 2, 1, 1, 0, 0, 3, 3, 1, 4, 3

Mr. Campbell made the following line plot of the data.

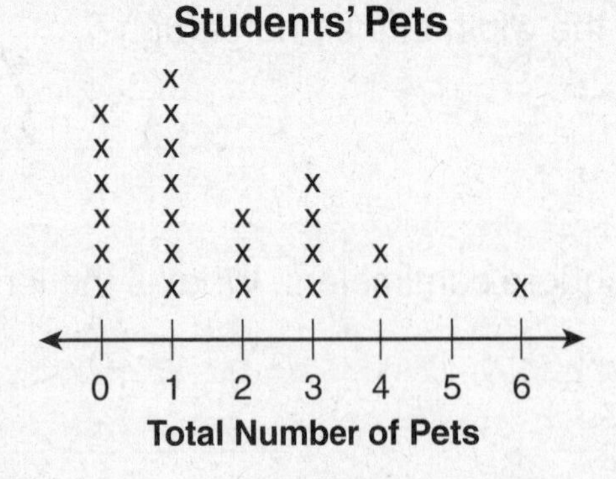

How many students in Mr. Campbell's class have 3 pets?
What is the greatest number of pets that any student in the class has?
What number of pets occurs most often in the data?
How does the line plot help you answer each of these questions?

To find the number of students who have a total of 3 pets, find the 3 on the number line. Count the number of X marks that are above the 3.
The line plot shows 4 X marks above the 3 on the number line.
So there are 4 students in Mr. Campbell's class who have 3 pets.

To find the greatest number of pets that any student in the class has, find the greatest number shown on the number line that has at least 1 X mark above it.
The greatest number on the number line that has at least 1 X mark above it is 6. So the greatest number of pets that any student in the class has is 6.

To find the number of pets that occurs most often in the data, find the number on the number line that has the greatest number of X marks above it.
The number on the number line with the most X marks above it is 1.
So the number of pets that occurs most often in the data is 1.

A line plot can also show a set of measurements in fractions of a unit.

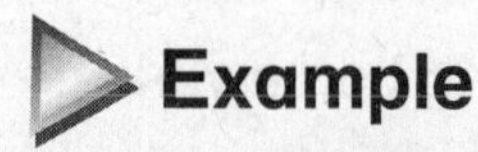

Example

The owner of a nursery measured the growth of several trees from April 1 to April 7. Here are the results in fractions of an inch:

$$\frac{1}{2}, \frac{3}{4}, \frac{1}{8}, \frac{3}{4}, 1, \frac{3}{8}, \frac{1}{4}, \frac{5}{8}, \frac{1}{4}, \frac{1}{2}, \frac{7}{8}, \frac{3}{4}, 1, \frac{3}{4}, \frac{1}{8}, \frac{1}{8}$$

Make a line plot of the data.

First, rename the measurements so that they are in the same fractional units. You can use eighths.

$$\frac{4}{8}, \frac{6}{8}, \frac{1}{8}, \frac{6}{8}, \frac{8}{8}, \frac{3}{8}, \frac{2}{8}, \frac{5}{8}, \frac{2}{8}, \frac{4}{8}, \frac{7}{8}, \frac{6}{8}, \frac{8}{8}, \frac{6}{8}, \frac{1}{8}, \frac{1}{8}$$

Then draw a number line. Use the measurements to label the tick marks. Label the number line to show what the measurements represent.

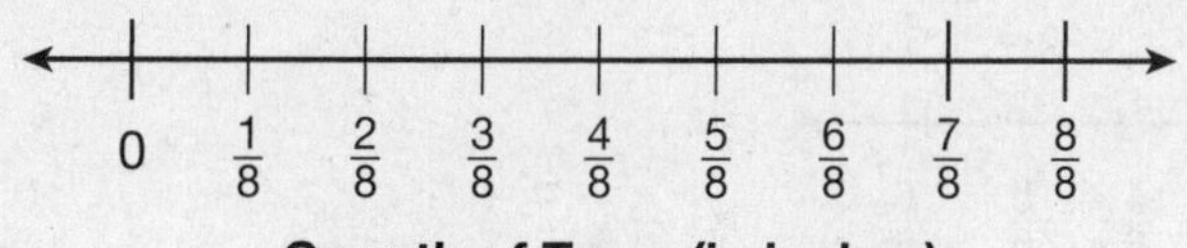

Finally, draw Xs to represent the data. Put one X above the measurement for each piece of data. Give the line plot a title.

Tree Growth April 1–April 7

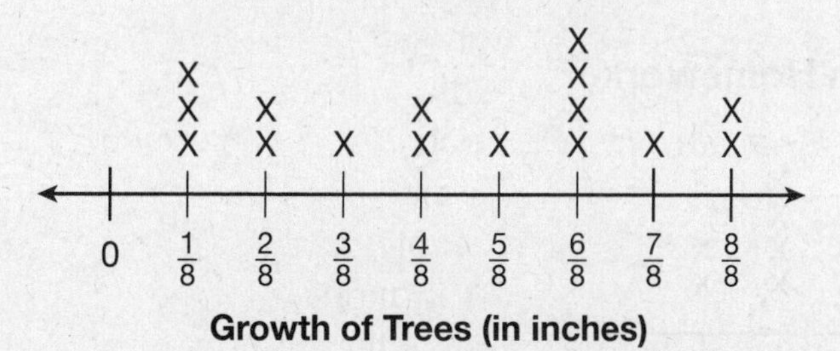

The data from a line plot can be used to solve word problems based on the data.

Example

Use the line plot above to answer this question: How many more trees grew $\frac{3}{4}$ of an inch than grew $\frac{3}{8}$ of an inch?

The line plot shows 1 X mark above the $\frac{3}{8}$-inch measurement and 4 X marks above the $\frac{6}{8}$-inch, or $\frac{3}{4}$-inch, measurement. There are 3 more X marks above the $\frac{6}{8}$ than there are above the $\frac{3}{8}$.

So 3 more trees grew $\frac{3}{4}$ of an inch than grew $\frac{3}{8}$ of an inch.

Example

Brad asked several of his classmates how long they spent doing homework last night. Here are the results in fractions of an hour:

$\frac{5}{6}, \frac{1}{3}, \frac{1}{2}, \frac{1}{6}, \frac{2}{3}, \frac{1}{3}, \frac{1}{2}, \frac{1}{6}, \frac{2}{3}, \frac{5}{6}, \frac{2}{3}, \frac{2}{3}, \frac{1}{2}, \frac{1}{2}, \frac{1}{2}$

Make a line plot for the data.

First, rename the measurements so they are in the same fractional units. You can use sixths.

$\frac{5}{6}, \frac{2}{6}, \frac{3}{6}, \frac{1}{6}, \frac{4}{6}, \frac{2}{6}, \frac{3}{6}, \frac{1}{6}, \frac{4}{6}, \frac{5}{6}, \frac{4}{6}, \frac{4}{6}, \frac{3}{6}, \frac{3}{6}, \frac{3}{6}$

Then draw a number line. Use the measurements to label the tick marks, and give the measurements a label.

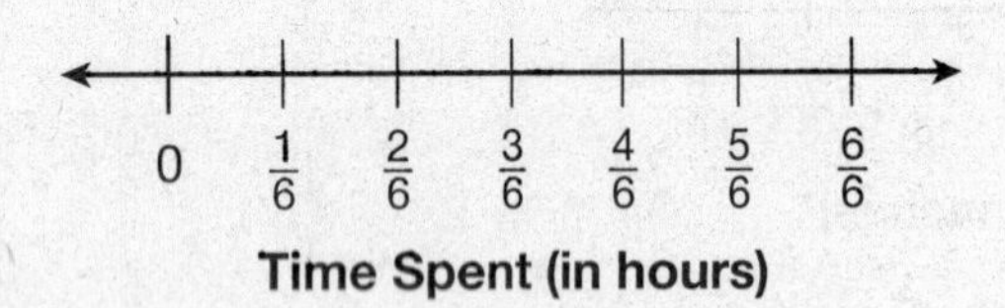

Next, draw Xs to represent the data.

Finally, give the line plot a title.

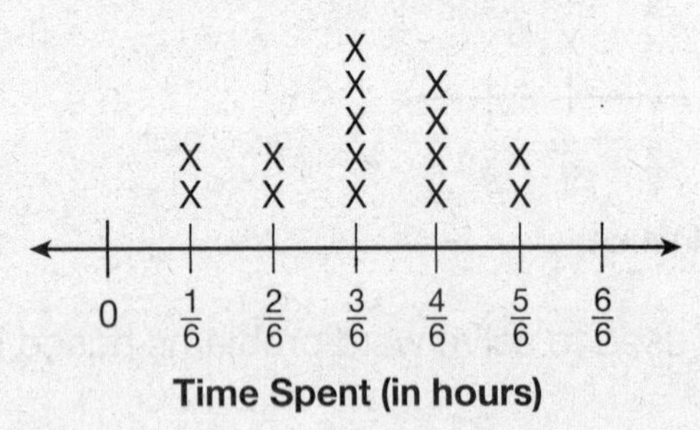

CCSS: 4.MD.4

Practice

Directions: Use the following information to answer questions 1 through 6.

The list below shows the distances in fractions of a mile that Tamara jogged last month.

$\frac{1}{2}, \frac{3}{4}, \frac{1}{4}, \frac{7}{8}, \frac{3}{8}, \frac{1}{2}, \frac{1}{4}, \frac{1}{2}, \frac{5}{8}, \frac{7}{8}, \frac{1}{2}, \frac{3}{4}, \frac{3}{4}, \frac{7}{8}, \frac{1}{8}, \frac{1}{2}, \frac{3}{4}$

1. Make a line plot of the set of data.

2. How many times last month did Tamara jog $\frac{3}{8}$ mile? ______________

3. What distance did Tamara jog twice last month? ______________

4. How many times did Tamara go jogging last month? ______________

5. What distance did Tamara jog the greatest number of times? ______________

6. How many more times did Tamara jog $\frac{1}{2}$ mile than $\frac{1}{8}$ mile? ______________

Directions: The line plot below shows the results of a survey of the students in Mrs. Hubbard's fourth-grade class. Use the line plot to answer questions 7 through 11.

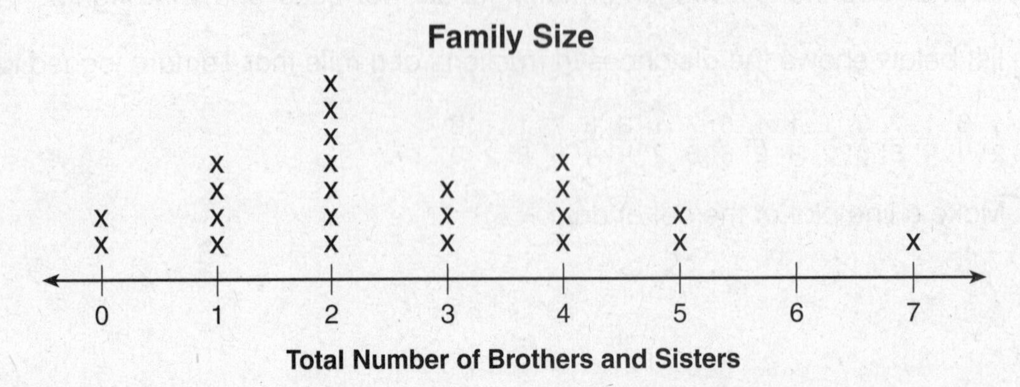

7. List the set of data from which this line plot could have been made.

____

____

____

8. How many students have a total of 4 brothers and sisters? ____________

9. What is the greatest total number of brothers and sisters of any student in Mrs. Hubbard's class?

____

10. How many students have a total of 2 brothers and sisters? ____________

11. How many students have a total number of brothers and sisters that is greater than 2?

____

CCSS: 4.MD.4

Directions: Carla measured the lengths of several insects in fractions of an inch. Then she made the line plot below to show the results. Use the line plot to answer questions 12 through 16.

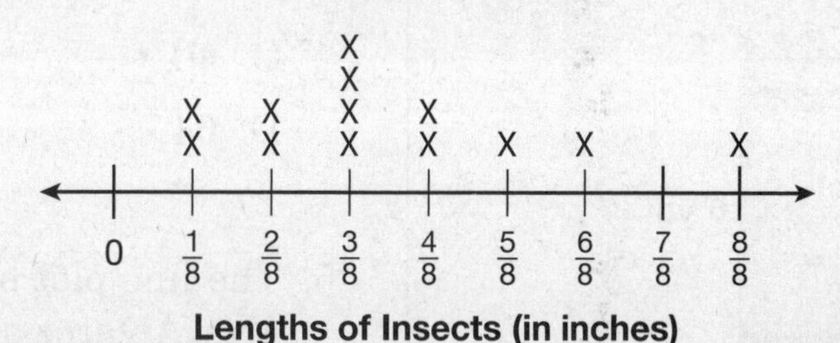

12. List the data.

____

____

____

13. How many insects are $\frac{3}{8}$ inch long? ____________

14. How many insects are less than $\frac{1}{2}$ inch long? ____________

15. How many insects are more than $\frac{3}{4}$ inch long? ____________

16. How many insects are $\frac{7}{8}$ inch long? How do you know?

____

Unit 4 Practice Test

Choose the correct answer.

1. What is the area of this rug?

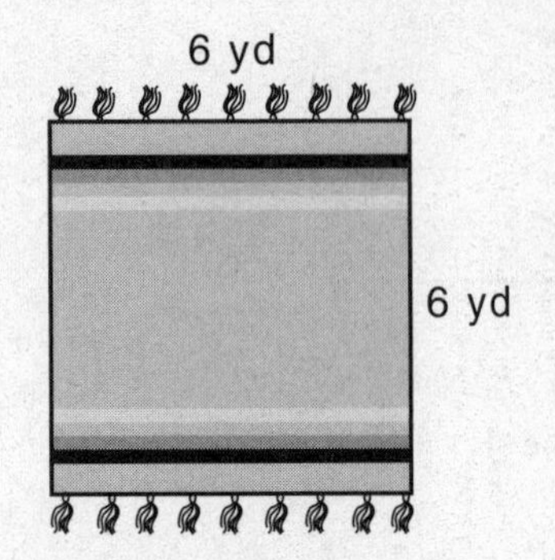

A. 12 square yards

B. 24 square yards

C. 30 square yards

D. 36 square yards

2. Which of the following units would you need the **fewest** of to measure the distance from New York City to Washington, D. C.?

A. inches

B. yards

C. miles

D. feet

3. George's dog has a mass of 4 kilograms. What is the mass of George's dog in grams?

A. 40

B. 400

C. 4,000

D. 40,000

4. How many cups are in 32 pints?

A. 8

B. 16

C. 40

D. 64

5. The line plot below shows the time Mrs. Alvarez spent cooking dinner each night for a month.

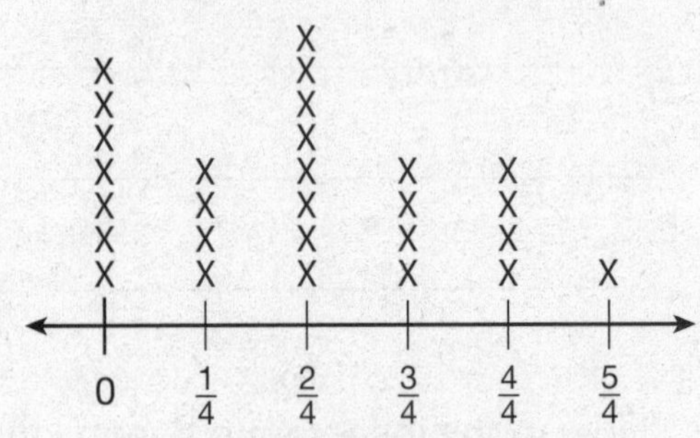

How many times did Mrs. Alvarez spend **more than** $\frac{3}{4}$ of an hour cooking dinner?

A. 5

B. 6

C. 7

D. 8

6. Mercedes bought a 3-pound package of rice. She used 12 ounces of rice on Monday and 10 ounces of rice on Tuesday. How many ounces of rice were left in the package then?

A. 26 ounces

B. 36 ounces

C. 38 ounces

D. 48 ounces

7. David walked 1 mile from his house to Frank's house. He stopped at the bookstore after walking for $\frac{3}{10}$ of a mile. Use the number line below to determine how much farther he had to go after stopping at the bookstore.

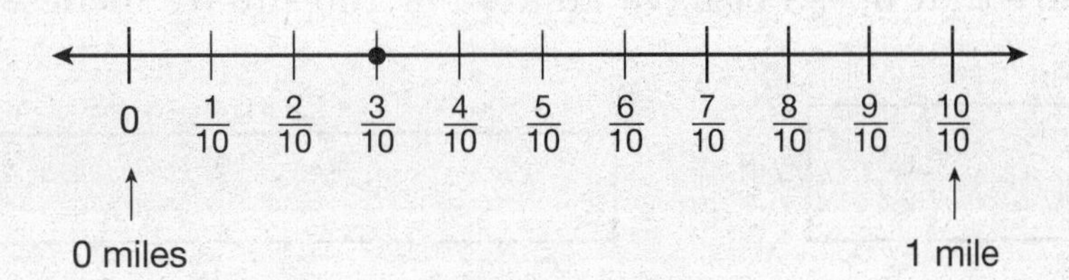

David still had ______________ of a mile to go.

8. A baseball game lasts about 3 hours, or ______________ minutes.

9. What is the perimeter of this rectangle?

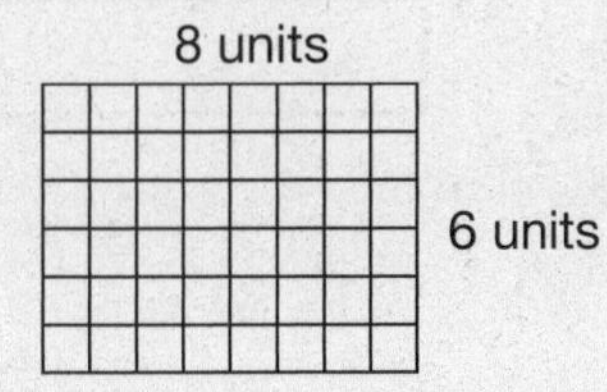

$P =$ ______________

10. Would it take a **greater** number of cups or quarts to measure the capacity of a kitchen sink? Explain your answer.

____

____

____

11. 5 lb = ____________ oz

12. Circle the figure that does not have an area of 100 square inches.

20 in.
5 in.

25 in.
4 in.

13. Denzel wrote (1, 12) to show that 1 foot is equal to 12 inches. What should he write to show how many inches 5 feet are equal to?

____

14. Stacey made a rectangular flower garden in her yard. The perimeter of the garden is 32 meters. Draw and label one possible garden that Stacey could have made.

Solve each problem.

15. Judy rode her bike 4 miles on Saturday. She rode $\frac{1}{2}$ of that distance on Sunday. What distance did Judy ride her bike on Sunday? ______________

16. The area of a rectangle is 40 square feet. The width of the rectangle is 5 feet. What is the length of the rectangle? ______________

17. Walter bought 3 bags of apples. Each bag had a mass of $\frac{3}{10}$ of a kilogram. What was the total mass of the apples that Walter bought? ______________

18. Nadeem bought 3 sweaters. The first sweater cost $29. The second sweater cost $8 less than the first sweater. The third sweater cost $6 more than the second sweater. How much did the third sweater cost? ______________

 Explain how you got your answer.

 ____

 ____

 ____

19. The gas tank in Jon's car holds 15 gallons of gas. How many quarts of gas does Jon's gas tank hold? ____________

20. Ines spent 3 hours gardening. She spent 20 minutes weeding and 20 minutes watering. She spent the rest of the time planting seeds. How many minutes did Ines spend planting seeds? ____________

21. A boulder in Nigel's front yard has a mass of 165 kilograms. What is the mass of the boulder in grams? ____________

22. Use a decimal to write 8 centimeters as a number of meters. ____________

23. Ricky and his twin brother Micky each measured their height in centimeters. Ricky is 142.39 cm tall, and Micky is 142.93 cm tall. Who is taller? ____________

Explain how you got your answer.

____

____

____

24. **Part A**
Use a ruler to measure the dimensions of this rectangle in centimeters.

length = ______________

width = ______________

Part B
What formula can you use to find the perimeter of the rectangle?

____

Use the formula and your measurements to find the perimeter of the rectangle.

P = ______________

Part C
What formula can you use to find the area of the rectangle?

____

Use the formula and your measurements to find the area of the rectangle.

A = ______________

25. A health-food store sells different-size bottles of vitamin water. Bottles hold from $\frac{1}{6}$ of a pint to $\frac{5}{6}$ of a pint. The list below shows the capacities of the bottles of vitamin water that were sold last week.

$\frac{1}{6}$, $\frac{1}{2}$, $\frac{1}{3}$, $\frac{1}{2}$, $\frac{5}{6}$, $\frac{2}{3}$, $\frac{1}{6}$, $\frac{1}{3}$, $\frac{1}{2}$, $\frac{2}{3}$, $\frac{5}{6}$, $\frac{2}{3}$, $\frac{2}{3}$, $\frac{1}{2}$, $\frac{5}{6}$, $\frac{1}{6}$

Part A

Rename the measurements so that they are in the same fractional units.

____

Part B

Make a line plot of the set of data.

Part C

How many bottles of vitamin water did the health-food store sell last week?

____

Part D

The manager of the health-food store is ready to order some more bottles of vitamin water from the factory. What size bottle (or bottles) should she order the most of? Explain your answer.

____

____

____

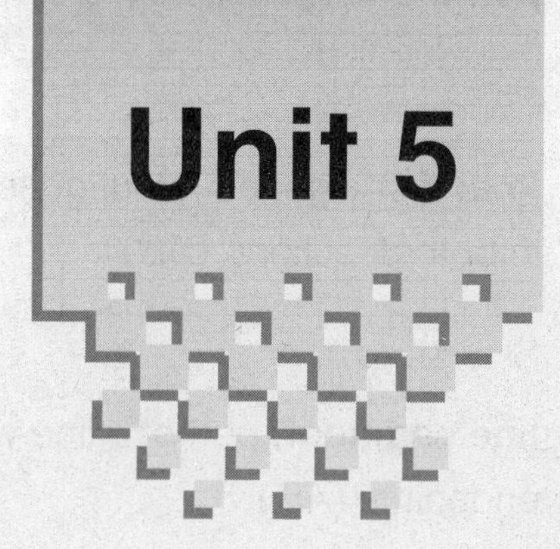

Geometry

Have you ever played in a tree house? Did you help build it? When you build a tree house, you are using geometry. You are also using geometry skills when you make a kite or figure out the best way to get to your friend's house. From the invention of the wheel to the exploration of our solar system, geometry has been helping people learn about and live in our world.

In this unit, you will learn about things that are all around you, including angles, lines, and polygons. You will also learn about symmetric figures and lines of symmetry.

In This Unit

Lesson 34: Points, Lines, and Planes

Points, lines, and planes are the basic building blocks of geometric figures.

point: a single location or position (*R*)

R

line: a straight path that goes on forever in both directions ($\overleftrightarrow{AB}$)

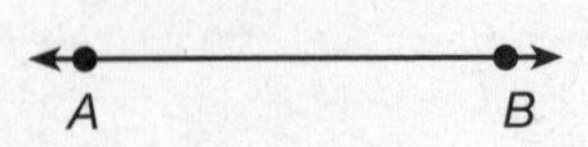

line segment: part of a line with two endpoints ($\overline{AB}$)

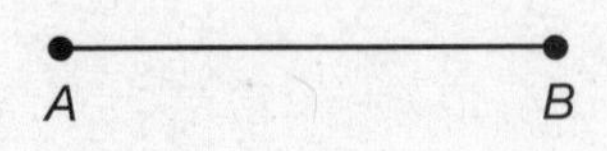

ray: part of a line that begins at one endpoint and goes on forever in one direction; when naming a ray, the letter of the endpoint is written first ($\overrightarrow{AB}$)

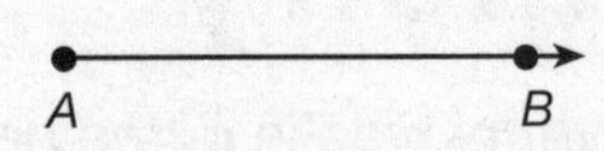

plane: a flat surface that goes on forever in all directions

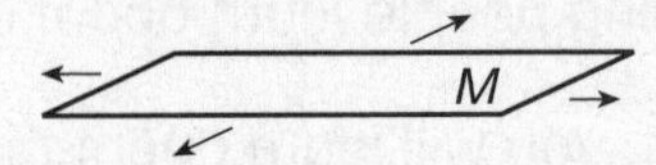

intersecting lines: lines that meet at a point

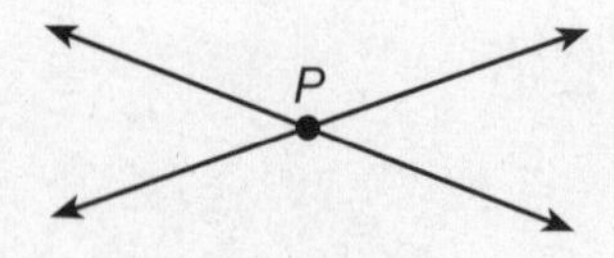

parallel lines: lines that are in the same plane (flat surface) and do not intersect

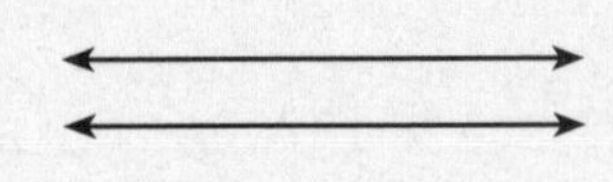

perpendicular lines: lines that intersect and form right angles (square corners); for example, the bottom corner of this page is a right angle formed by the side and bottom edges of the page

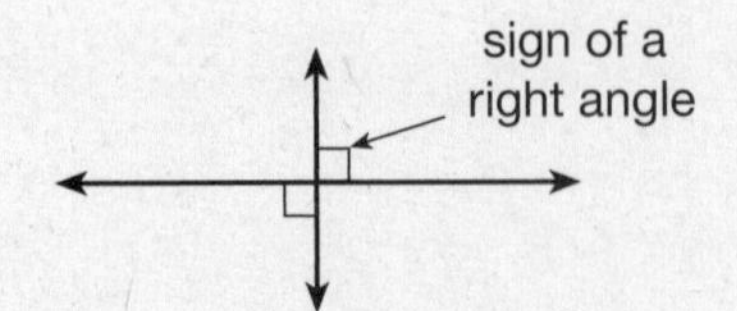

Practice

Directions: For questions 1 and 2, label the type of lines that each real-world object represents (parallel, intersecting, or perpendicular).

1. spokes of tire

2. angle made by sign and post

3. Find two objects in your classroom and name the types of lines they have.

	objects	lines
A.		
B.		

4. Write *parallel lines* or *perpendicular lines* on the blank under each drawing.

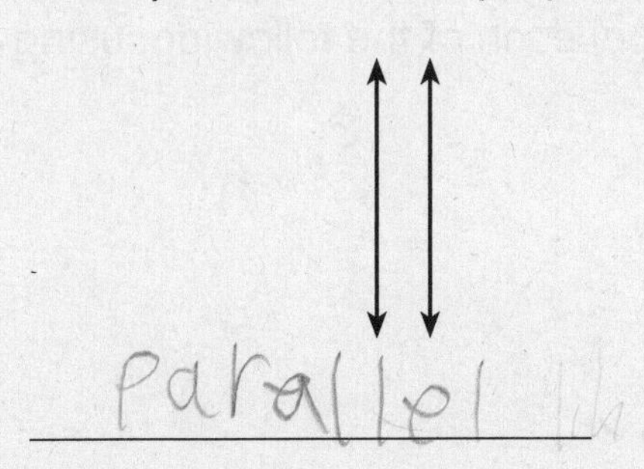

5. What is shown in the figure below?

A. point
B. line
C. line segment
D. ray

6. What is shown in the figure below?

A. point
B. line
C. line segment
D. ray

Directions: Use the following figures to complete questions 7 through 11.

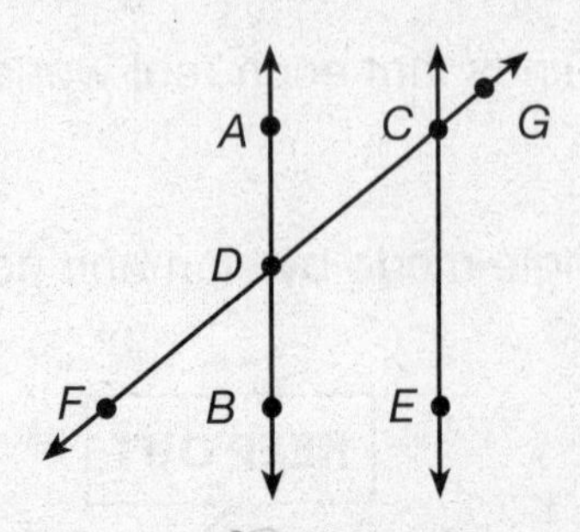

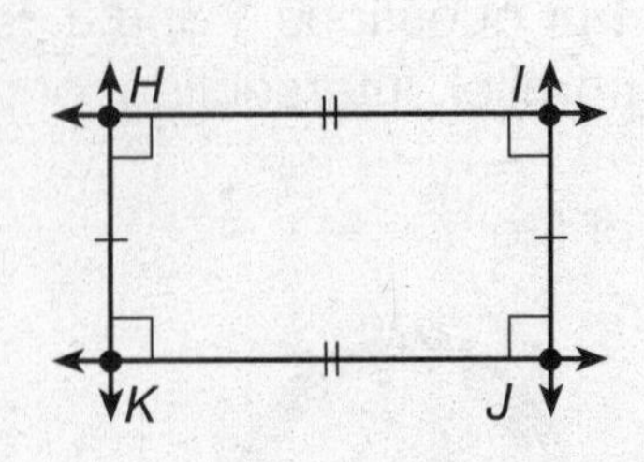

7. $\overleftrightarrow{AB}$ and $\overleftrightarrow{FG}$ are Intersecting lines.

8. $\overleftrightarrow{AB}$ and $\overleftrightarrow{CE}$ are Parallel lines.

9. $\overrightarrow{FG}$ is a line.

10. $\overline{DC}$ is a line segment.

11. Name four pairs of perpendicular lines. IJ, HIK, HK, HI

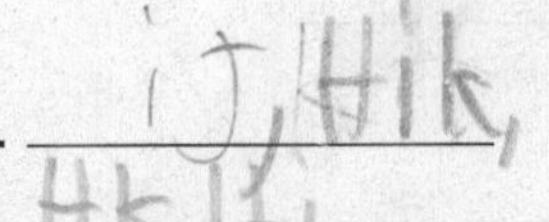

Directions: For questions 12 through 17, draw and label each of the following, using a straightedge or ruler if necessary.

12. $\overleftrightarrow{AB}$

13. $\overrightarrow{AB}$

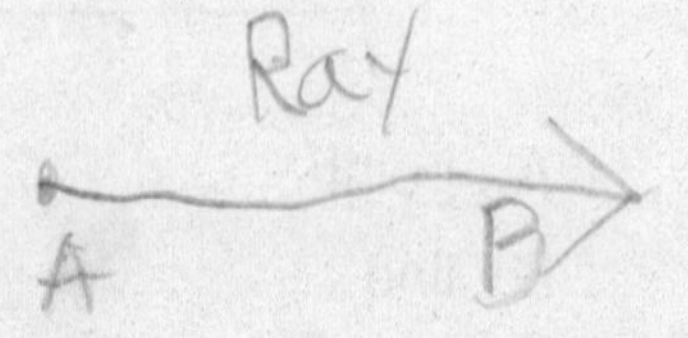

14. $\overline{AB}$

15. $\overleftrightarrow{AB}$ is parallel to $\overleftrightarrow{YZ}$

16. $\overleftrightarrow{AB}$ intersects $\overleftrightarrow{YZ}$

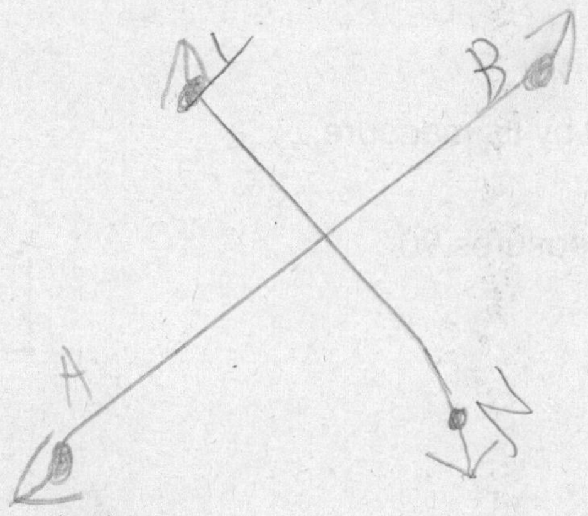

17. $\overleftrightarrow{AB}$ is perpendicular to $\overleftrightarrow{YZ}$

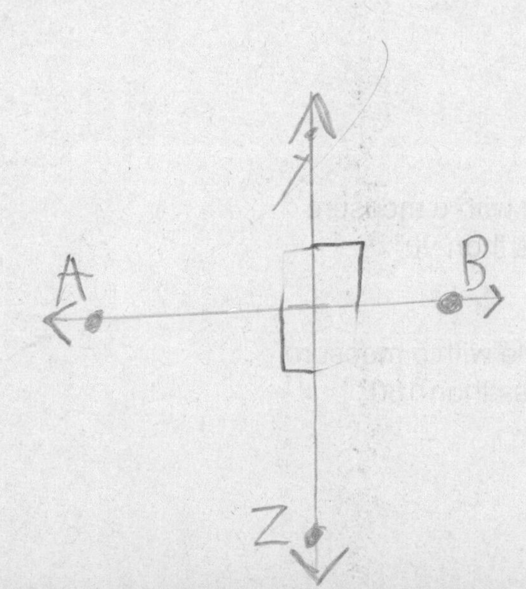

Lesson 35: Angles

An **angle** is formed by two **rays** with a common endpoint. The common endpoint is called the **vertex**. An angle is named by a point on one ray, the vertex, and a point on the other ray. You can also name an angle by using just the vertex point if that is the only angle using that point as its vertex.

Example

Write the name of this angle. Then, identify the vertex.

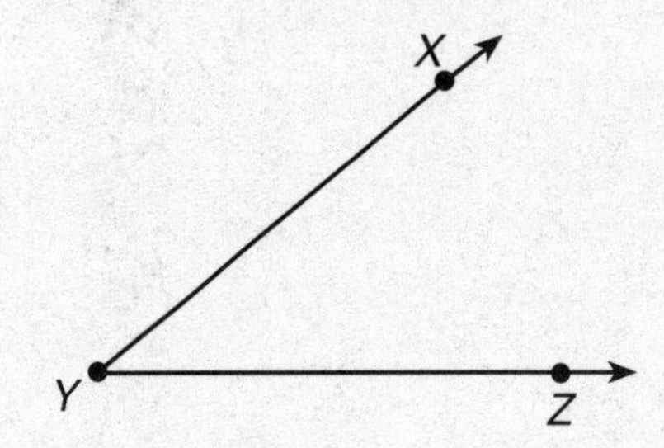

name: angle *XYZ*, angle *ZYX*, or angle *Y*
vertex: *Y*
rays: $\overrightarrow{YX}$ and $\overrightarrow{YZ}$

An angle's type is determined by its measure.

Right angle: an angle that measures 90°

Straight angle: an angle that measures 180°

Acute angle: an angle with a measure greater than 0° but less than 90°

Obtuse angle: an angle with a measure greater than 90° but less than 180°

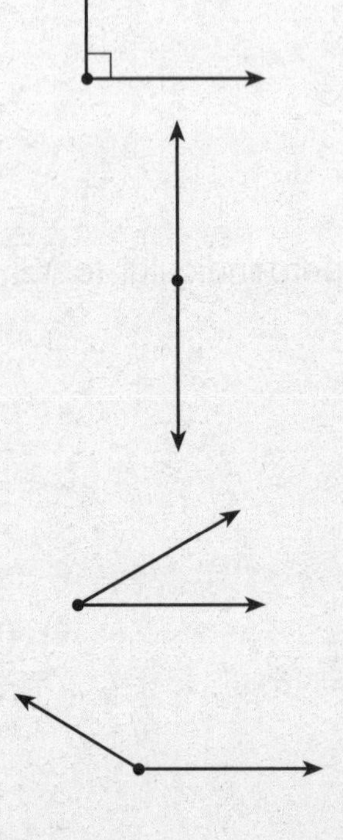

CCSS: 4.MD.5a,b; 4.MD.6, 4.MD.7, 4.G.1

The center of a circle can serve as the vertex for many different types of angles.

Example

This diagram shows an acute angle measuring 50° having its vertex at the center of a circle.

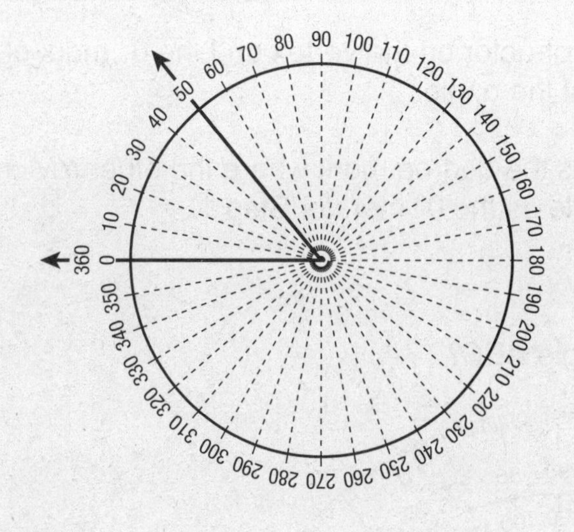

Example

This diagram shows an obtuse angle.

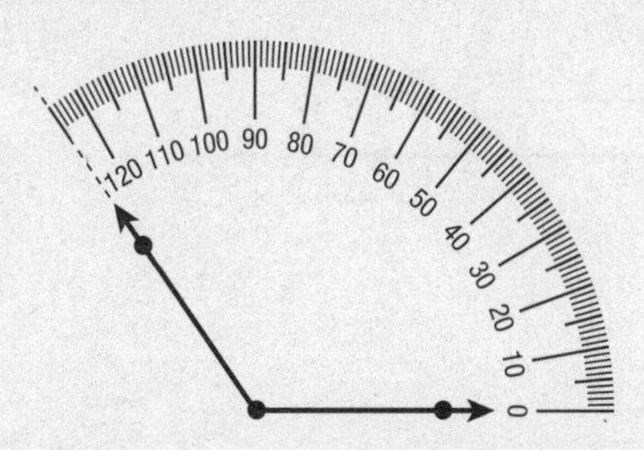

The angle's vertex is the center of a circle. One ray intersects the circle at the 125-degree mark. The other ray intersects the circle at the 0-degree mark. So the angle measures 125 degrees.

Each degree of the circle is $\frac{1}{360}$ of the whole circle, so the portion of the circle from one point of intersection to the other is $\frac{125}{360}$.

Each degree of an angle represents a turn of $\frac{1}{360}$ of a circle around the center of the circle. So the 125-degree angle represents a turn of $\frac{125}{360}$ of a circle, or 125×1 degree.

When you measure an angle, you are not finding the length of the angle's sides. You are measuring the **degree of rotation** of one of the angle's sides from its other side. Angles are measured in **degrees**.

You can use a **protractor** to measure an angle. When measuring an angle with a protractor, follow these steps:

Step 1: Line up the center of the protractor on the vertex and the 0° mark of one of the scales on one of the rays of the angle.

Step 2: The measure of the angle is the degree mark where the other ray crosses the protractor on the same scale as the 0° mark of Step 1.

Example

What is the measure of angle *ABC*?

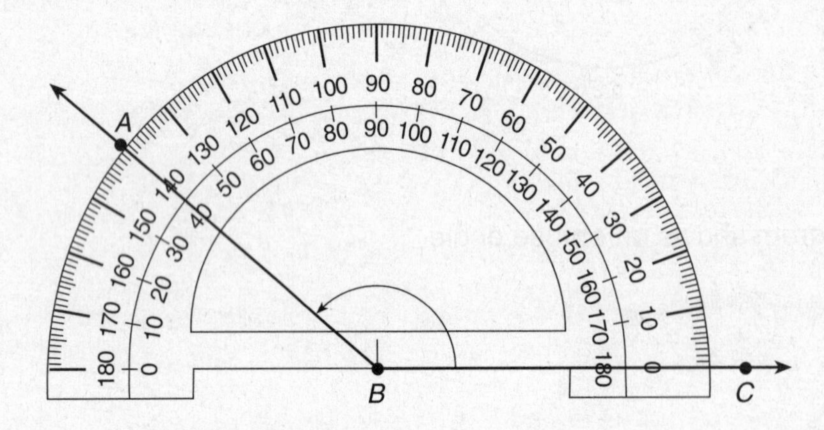

The measure of angle *ABC* is 140°.

TIP: Sometimes you may need to extend the rays of the angle so that you can see where the rays cross the marks on the protractor.

CCSS: 4.MD.5a,b; 4.MD.6, 4.MD.7, 4.G.1

To draw an angle with a given measure, follow these steps:

Step 1: Draw and label a ray.

Step 2: Line up the center of the protractor on the endpoint of the ray and the 0° mark of one of the scales on the ray.

Step 3: Mark and label a point where the given measurement is on the same scale as the 0° mark of Step 2.

Step 4: Use a straightedge to draw a ray from the endpoint of the first ray through the point that you marked in Step 3.

Example

Draw angle *JKL* with a measure of 35°.

Draw and label $\overrightarrow{KL}$. Notice that *K* will be the vertex of the angle.

Line up the center of the protractor on *K* and be sure $\overrightarrow{KL}$ crosses the protractor at the 0° mark. Mark and label *J* at 35° on the outside scale.

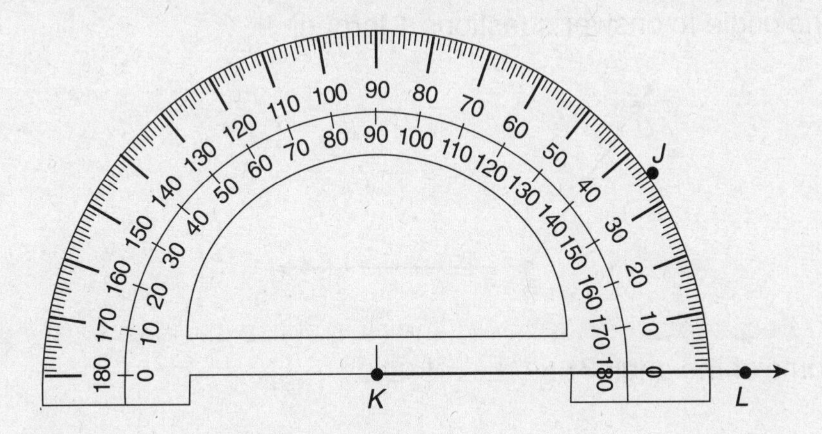

Use a straightedge to draw $\overrightarrow{KJ}$.

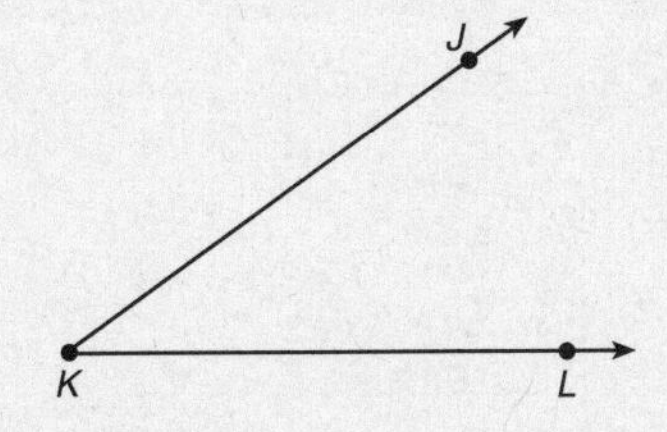

Sometimes a large angle is made up of two smaller angles. You can use the measure of the large angle and the measure of one of the smaller angles to find the measure of the other smaller angle.

Example

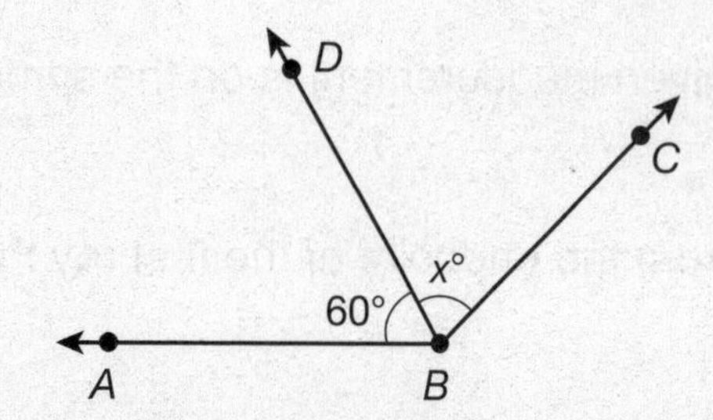

Angle *ABC* measures 135°, and angle *ABD* measures 60°. What is the measure of angle *DBC*?

$60 + x = 135$

$x = 75$

The measure of angle *DBC* is 75°.

Practice

Directions: Use the angle to answer questions 1 through 4.

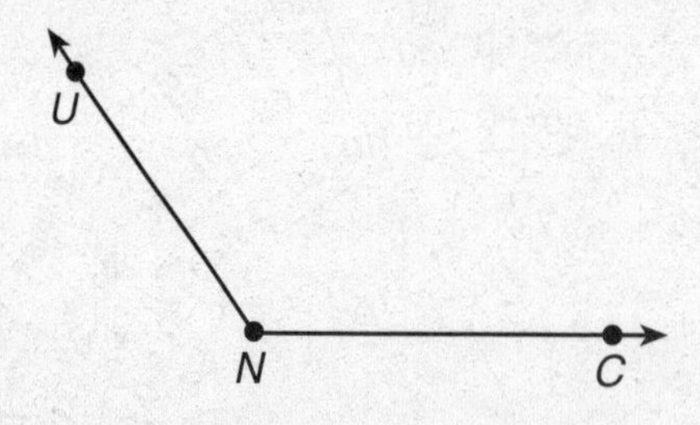

1. What is the name of the angle? UNC
2. What is the vertex? N
3. What are the rays of the angle? U/C
4. What type of angle is this? Obtuse

CCSS: 4.MD.5a,b; 4.MD.6, 4.MD.7, 4.G.1

5. What is the measure of angle *JKL*?

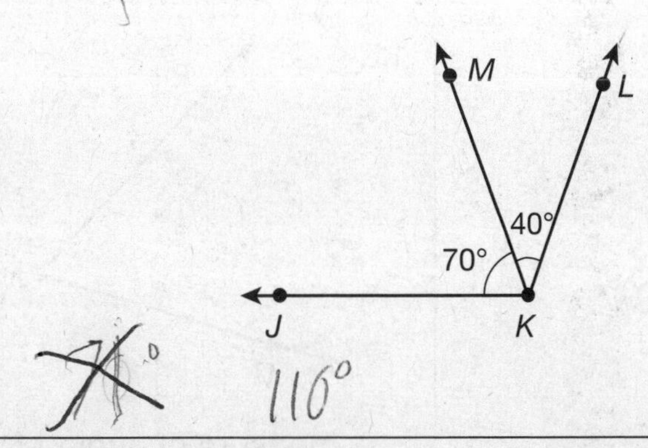

~~7°~~ 110°

Explain how you got your answer.

their was a little hole and I matched it with the dot in the middle and L lead to ~~x°~~ 110°

6. Angle *PQR* measures 150°, and angle *SQR* measures 100°. What is the measure of angle *PQS*?

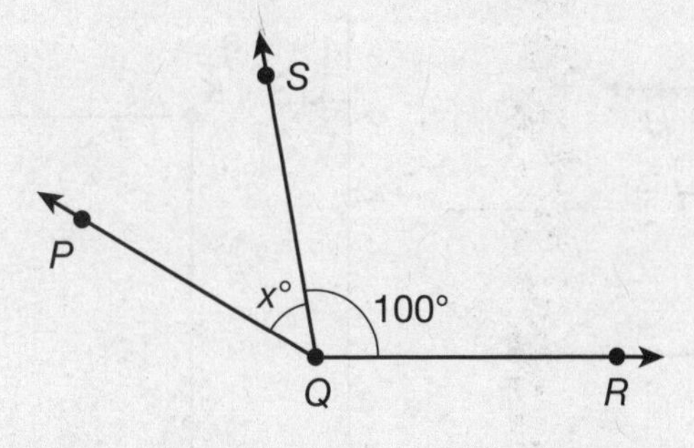

150°

Explain how you got your answer.

Since the arrow was Pointing to 130 and the angle was acute ~~So 50 is less then 130~~ So it is 150°

Directions: For questions 7 through 14, find the measure of the given angle.

7.

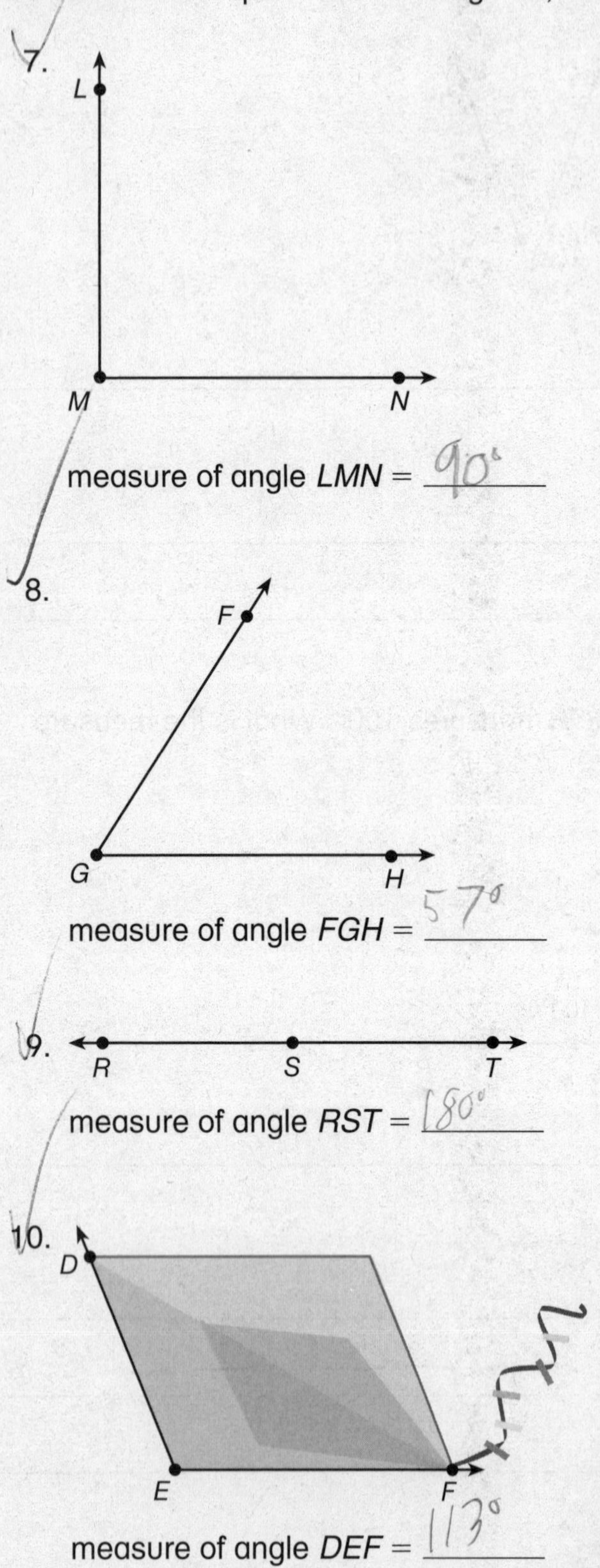

measure of angle *LMN* = ______

8.

measure of angle *FGH* = ______

9.

measure of angle *RST* = ______

10.

measure of angle *DEF* = ______

11.

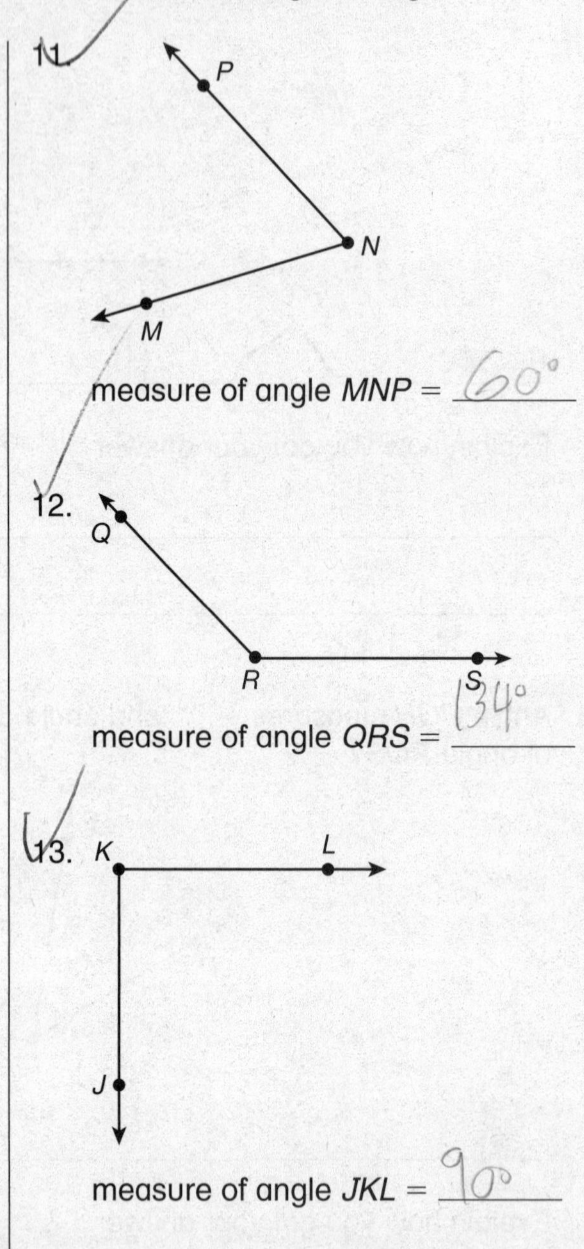

measure of angle *MNP* = ______

12.

measure of angle *QRS* = ______

13.

measure of angle *JKL* = ______

14.

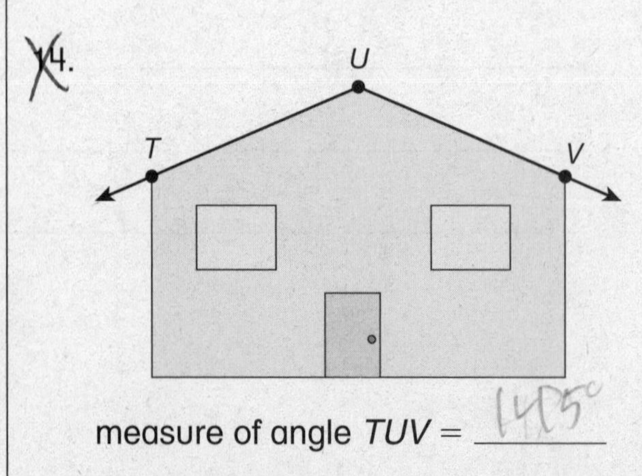

measure of angle *TUV* = ______

CCSS: 4.MD.5a,b; 4.MD.6, 4.MD.7, 4.G.1

Directions: For questions 15 through 18, draw an angle with the given name and measure. Then, name the vertex and the rays.

15. measure of angle $BCD = 135°$

vertex ____________

rays ____________

16. measure of angle $XYZ = 20°$

vertex ____________

rays ____________

17. measure of angle $LMN = 95°$

vertex ____________

rays ____________

18. measure of angle $RST = 70°$

vertex ____________

rays ____________

19. What type of angle does the following angle appear to be?

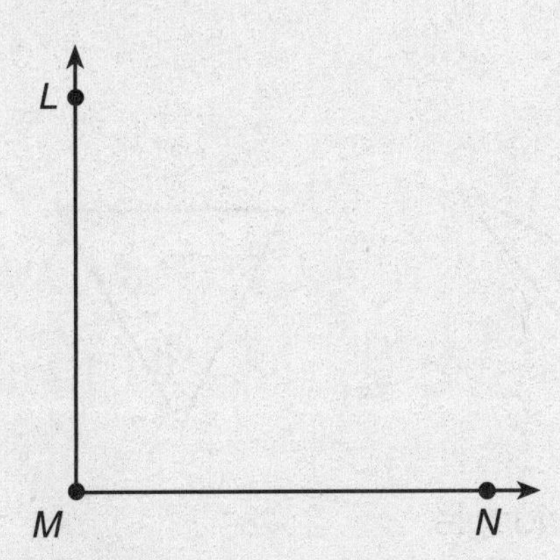

A. acute
B. right
C. straight
D. obtuse

20. Which of the following is an acute angle?

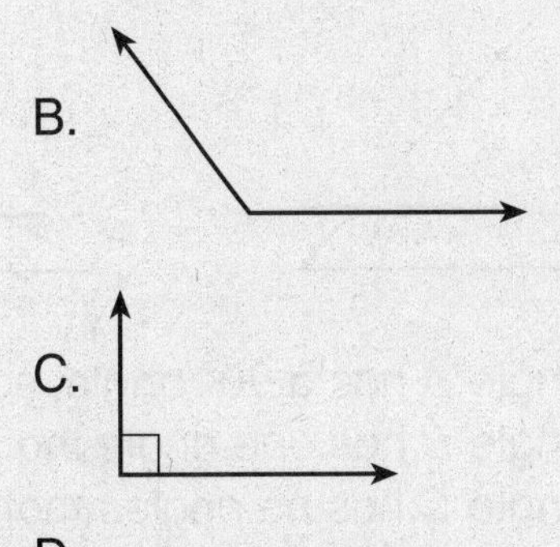

Lesson 36: Two-Dimensional Shapes

A **triangle** is a figure with three sides. Triangles can be named by the lengths of their sides.

Scalene Triangle	Isosceles Triangle	Equilateral Triangle
no equal sides or equal angles	at least 2 equal sides and at least 2 equal angles	3 equal sides and 3 equal angles

Triangles can also be named by the sizes of their angles. Every triangle has three angles. The sum of the measures of the three angles is 180°.

Right Triangle	Acute Triangle	Obtuse Triangle
acute, acute	acute, acute, acute	acute, acute, obtuse
has a right angle (measures 90°)	all angles measure less than 90°	has an angle that measures more than 90°

Example

Name each of these triangles by its angles.

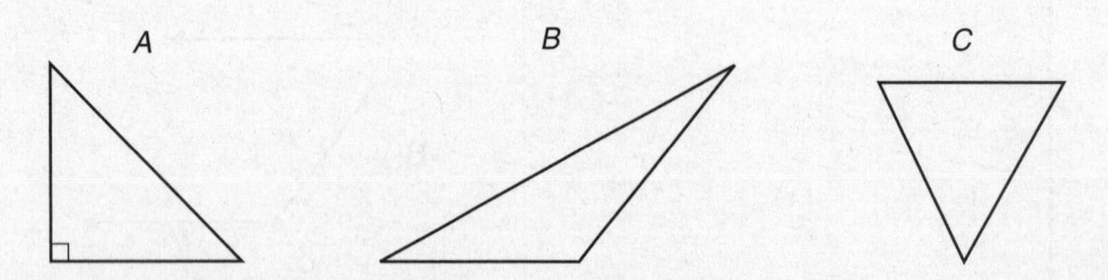

Triangle *A* has a 90° angle, so it is a right triangle.
Triangle *B* has one angle that is greater than 90°, so it is an obtuse triangle.
Triangle *C* has no angles that are 90° or greater, so it is an acute triangle.

CCSS: 4.G.1, 4.G.2

Polygons are closed plane figures made up of 3 or more straight sides. Polygons are named by the numbers of sides and angles they have. You already learned about polygons with 3 sides and angles (triangles) on the previous page. On this page, you will learn about more polygons. Polygons that have 4 sides and angles are quadrilaterals. Here are the descriptions and some examples of special quadrilaterals.

Parallelogram	Rhombus	Rectangle
opposite sides are congruent and parallel	a parallelogram with 4 congruent sides	a parallelogram with 4 right angles
Square	**Trapezoid**	**Isosceles Trapezoid**
a parallelogram with 4 congruent sides and 4 right angles	1 pair of parallel sides	1 pair of parallel sides and 1 pair of congruent sides

Other common polygons are the **pentagon**, **hexagon**, and **octagon**.

Pentagon	Hexagon	Octagon
a polygon with 5 sides and angles	a polygon with 6 sides and angles	a polygon with 8 sides and angles

TIP: Matching tick marks indicate congruent sides, and matching arcs indicate congruent angles.

Practice

Directions: For questions 1 through 4, classify the given triangle by angle measure using one of the following: right, acute, or obtuse. Then, classify the triangle by side length using one of the following: scalene, isosceles, or equilateral.

1.

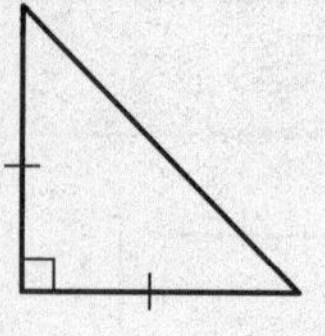

____

____

2.

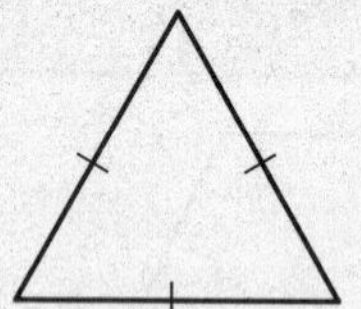

____

____

3.

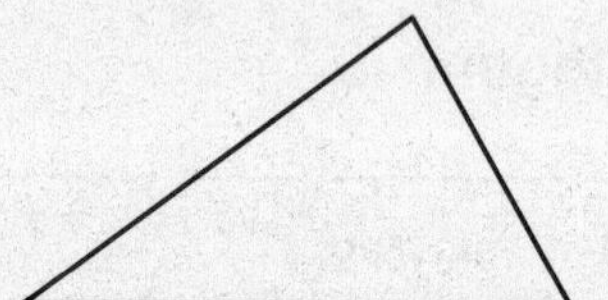

____

____

4.

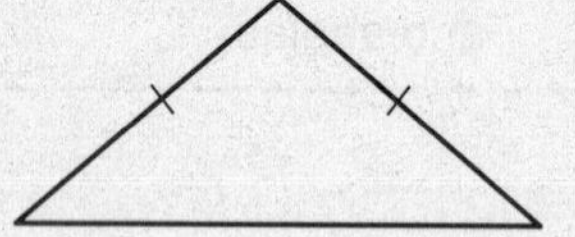

____

____

CCSS: 4.G.1, 4.G.2

Directions: For questions 5 through 11, use the given angle measurements to name the triangle.

5. 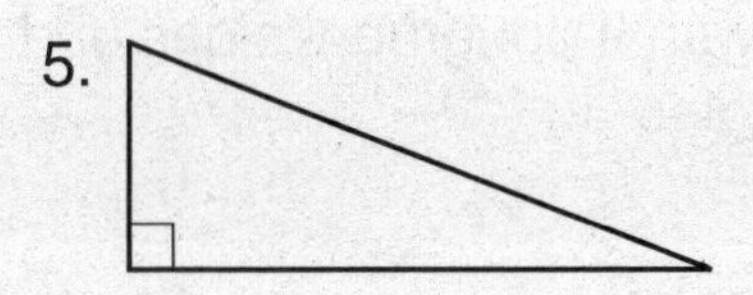

____

6. 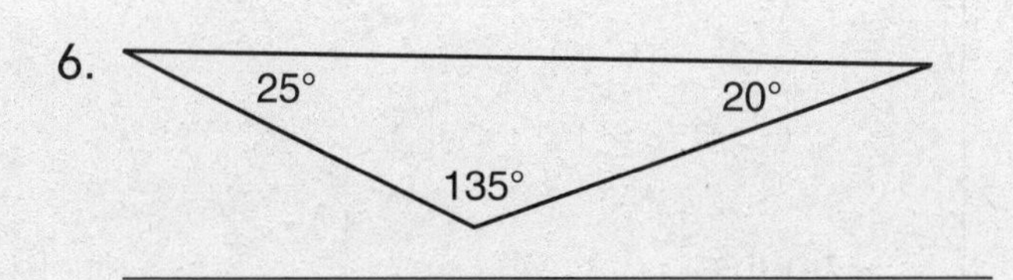

____

7. 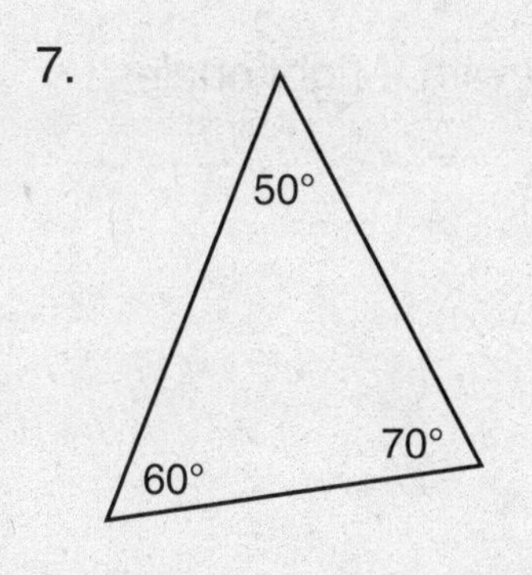

____

8. 60°, 60°, 60° ______________________________

9. 26°, 90°, 64° ______________________________

10. 30°, 30°, 120° ______________________________

11. 6 ft, 10 ft, 6 ft ______________________________

Directions: For questions 12 through 17, draw a figure for each description. Then write the name of the figure.

12. a polygon with 5 sides and angles

    figure ______________

13. a parallelogram with 4 equal sides and 4 right angles

    figure ______________

14. a quadrilateral with only 1 pair of parallel sides

    figure ______________

15. a polygon with 4 congruent sides and no right angles

    figure ______________

16. a quadrilateral with 4 right angles

    figure ______________

17. a polygon with 4 sides and angles

    figure ______________

Directions: Use the following figures to answer questions 18 through 23.

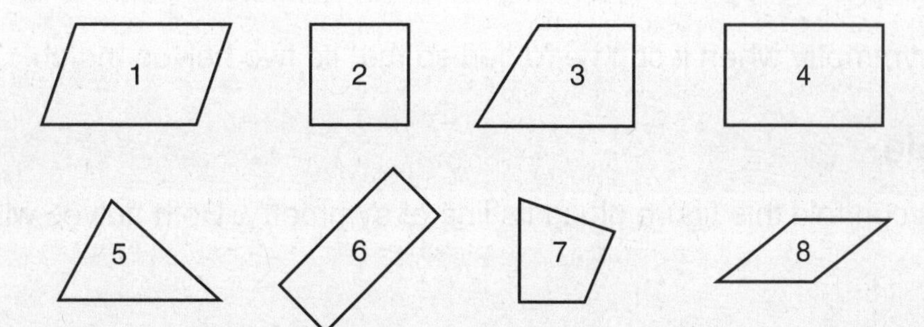

18. Write the numbers of all the figures that have two sets of parallel sides.

____

19. Write the numbers of all the figures that are rectangles.

____

20. Write the number of each figure that appears to be a rhombus.

____

21. Describe one way in which figures 1 and 3 are the same and one way they are different.

____

____

____

22. Which of the following best describes figure 5?

    A. right triangle
    B. scalene triangle
    C. isosceles triangle
    D. equilateral triangle

23. Which of the figures can be classified as a parallelogram?

    A. 1
    B. 1 and 8
    C. 1, 6, and 8
    D. 1, 2, 4, 6, and 8

Lesson 37: Symmetry

A figure has symmetry when it can be folded so that its two halves match.

Example

You can fold this figure along its line of symmetry. Both halves will match.

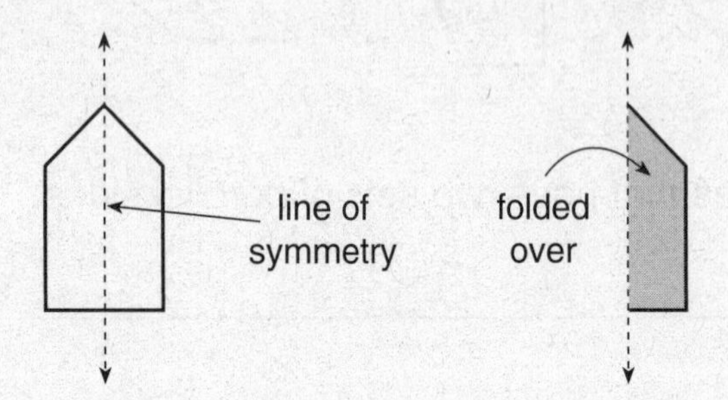

Some figures have more than one line of symmetry.

Example

The following figure has two lines of symmetry. It can be folded along either dotted line to form two equal halves.

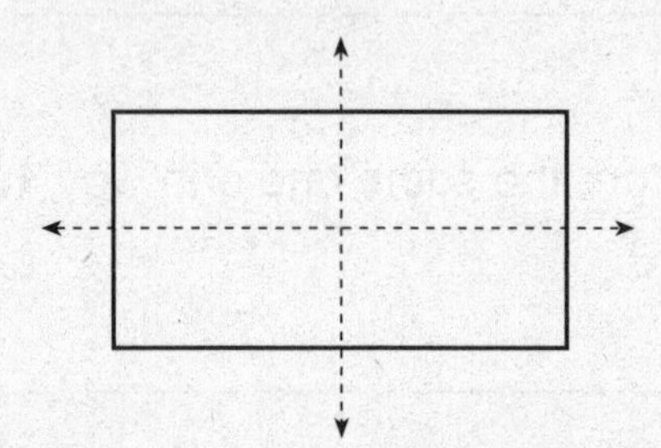

Some figures have no lines of symmetry.

Example

This figure has no lines of symmetry. There is no way to fold it so that two halves match.

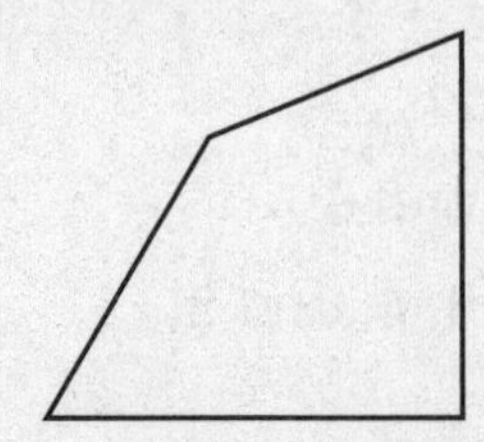

CCSS: 4.G.3

Practice

1. Half of this figure is missing. Using the line of symmetry, draw and shade the matching half of the figure.

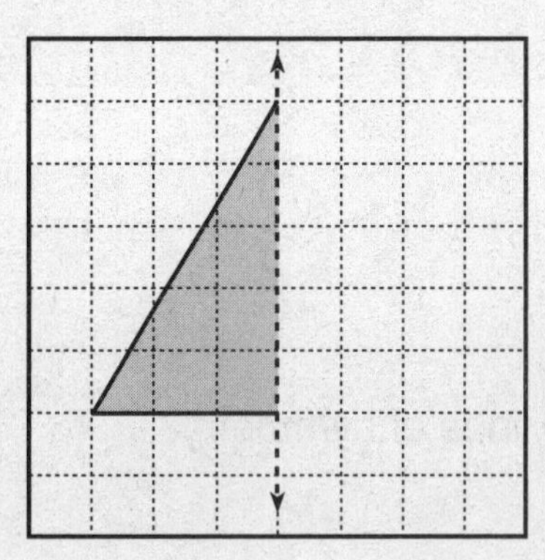

2. How many lines of symmetry does the following figure have? _______________

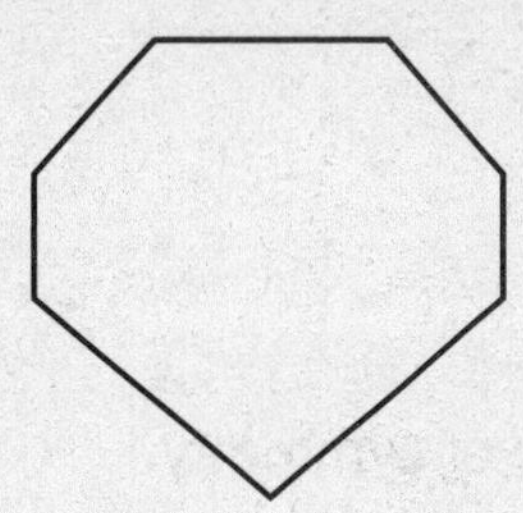

3. Circle the figure below that does **not** have a line of symmetry.

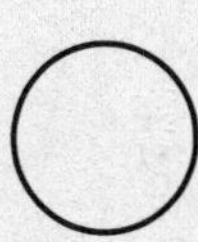 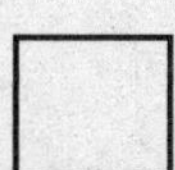  

4. Draw the line(s) of symmetry on each figure below.

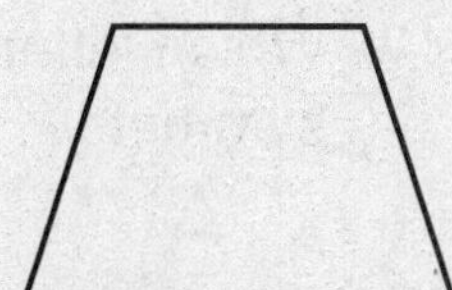 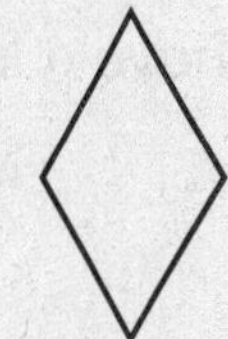

5. Draw a figure that has no lines of symmetry.

6. Draw a figure that has two lines of symmetry.

7. Draw a figure that has one line of symmetry.

8. Which letter does **not** have a line of symmetry?
   A. C
   B. Z
   C. W
   D. X

9. Which letter has two lines of symmetry?
   A. A
   B. D
   C. K
   D. X

10. How many lines of symmetry does a square have? ______________

Explain your thinking.

____

____

____

11. How many lines of symmetry does a parallelogram without any 90-degree angles have?

____

Explain your thinking.

____

____

____

12. How many lines of symmetry does an equilateral triangle have?

____

Explain your thinking.

____

____

____

13. How many lines of symmetry does a scalene right triangle have?

____

Explain your thinking.

____

____

____

Unit 5 Practice Test

Choose the correct answer.

1. Which type of triangle is shown?

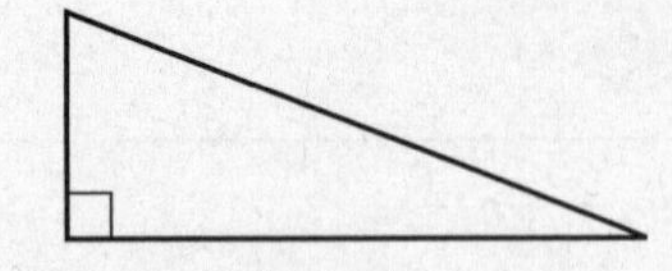

A. acute triangle
B. obtuse triangle
C. right triangle
D. equilateral triangle

2. Which figure shows a line of symmetry?

A.
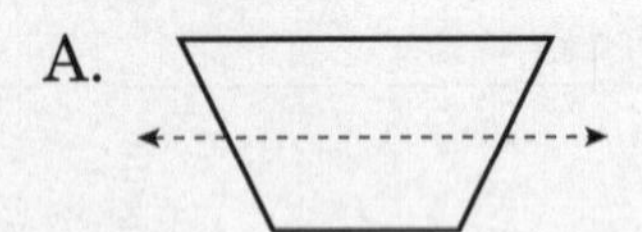

B.
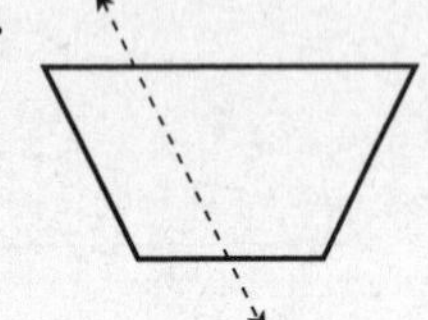

C.
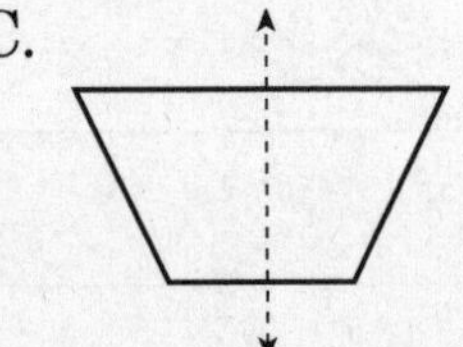

D.

3. Which of these is an obtuse angle?

A.
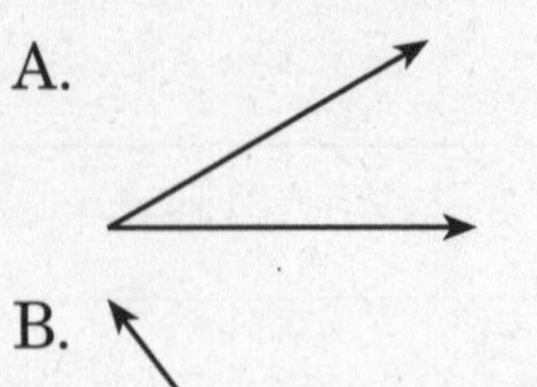

B.
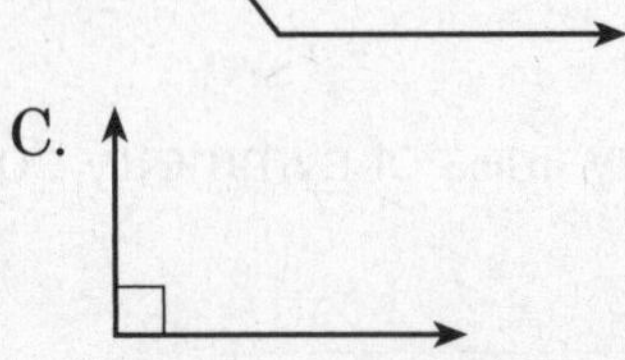

C.

D.

4. What geometric figure is shown?

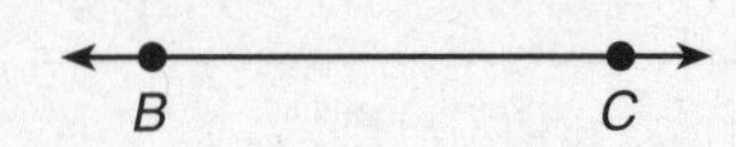

A. ray
B. line
C. line segment
D. point

5. Which streets are parallel in the map shown below?

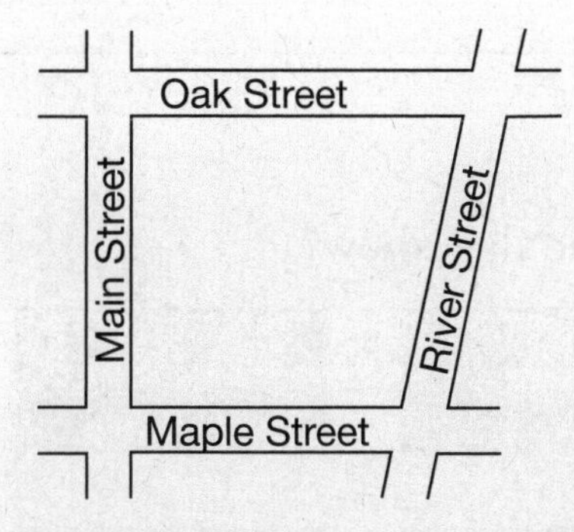

6. Use your protractor to draw angle *JKL* that measures 65°. Label points *J, K,* and *L* on your drawing.

7. List three quadrilaterals that are parallelograms.

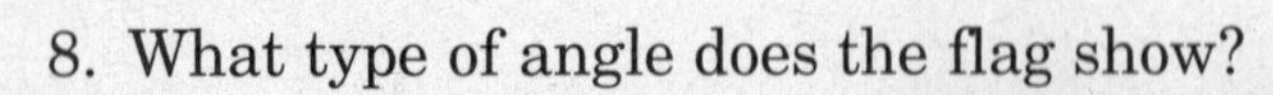

8. What type of angle does the flag show?

9. Rashid drew a triangle with angle measures of 40°, 65°, and 75°. What type of triangle did Rashid draw?

____

10. How many degrees are in an angle that turns through $\frac{1}{360}$ of a circle?

____

11. What is the measure of angle *DEF*?

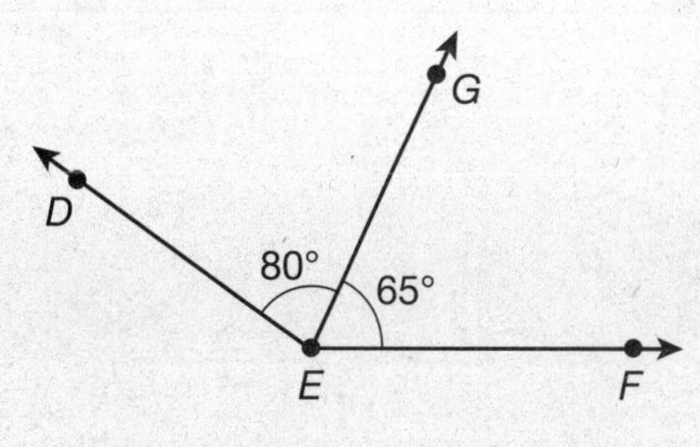

12. How many right angles does an isosceles trapezoid have?

13. Use your protractor to measure angle *QRS* below.

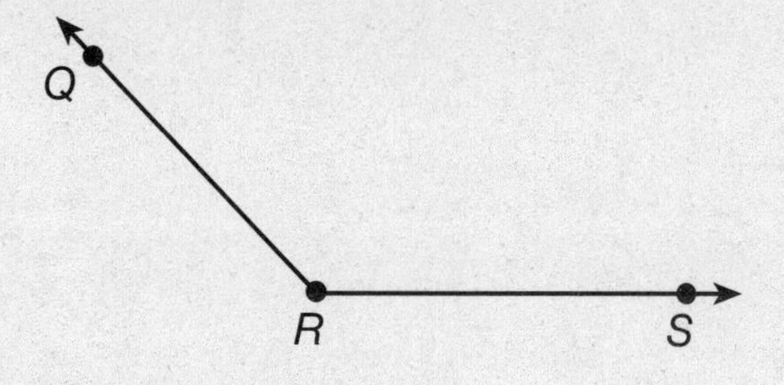

What is the measure of angle *QRS*?

14. Draw a square and one of its diagonals. Classify the triangles by their angles and by their sides.

15. Draw a figure that has one line of symmetry.

16. Angle *MNO* measures 60°, and angle *MNP* measures 28°. What is the measure of angle *PNO*?

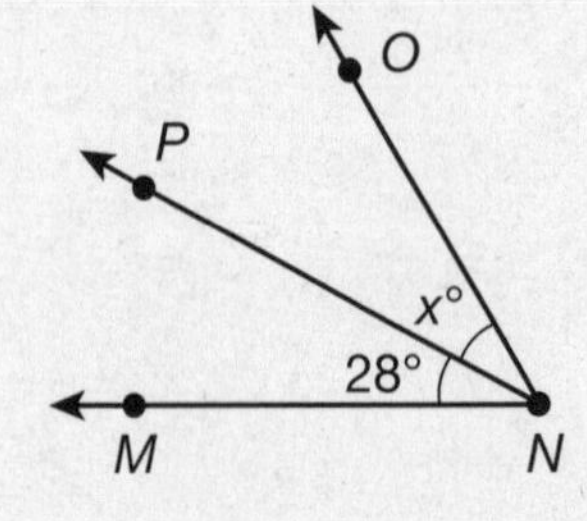

17. **Part A**
Draw quadrilateral *WXYZ* that is both a rectangle and a rhombus.

Part B
What is the name of this type of quadrilateral?

____

Part C
List two pairs of parallel sides.

____

____

18. **Part A**

Jake wrote his name in capital letters.

JAKE

Find the letters in Jake's name that have symmetry. Draw lines of symmetry on them.

How many letters in the name JAKE have symmetry?

____

Part B

Write your name in capital letters.

Find the letters in your name that have symmetry. Draw lines of symmetry on them.

How many letters in your name have symmetry?

____

Math Tool: Place-Value Models

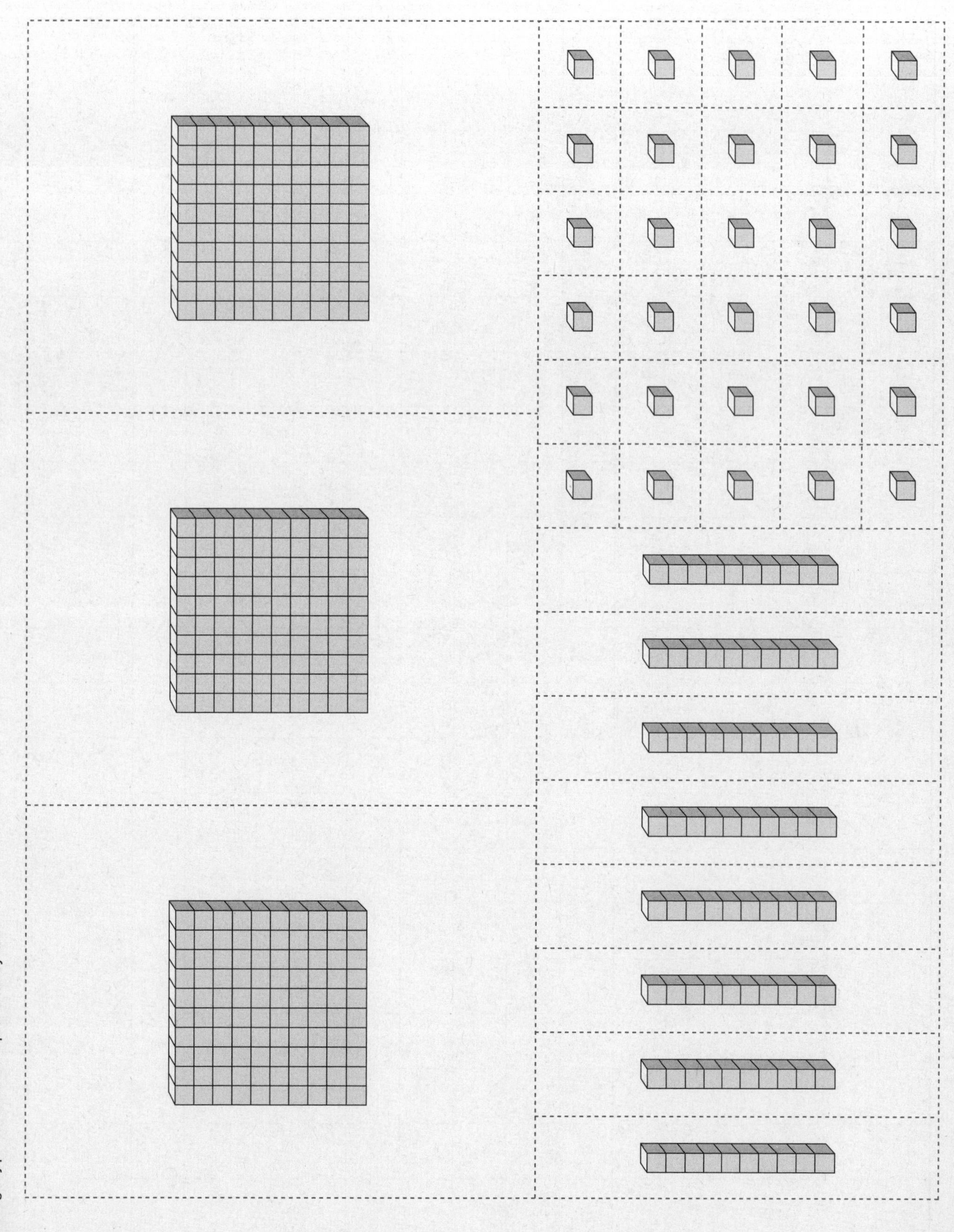

Cut or tear carefully along this line.

Math Tool: Grid Paper

Cut or tear carefully along this line.

Math Tool: Fraction Strips

Cut or tear carefully along this line.

Math Tool: Decimal Models for Tenths and Hundredths

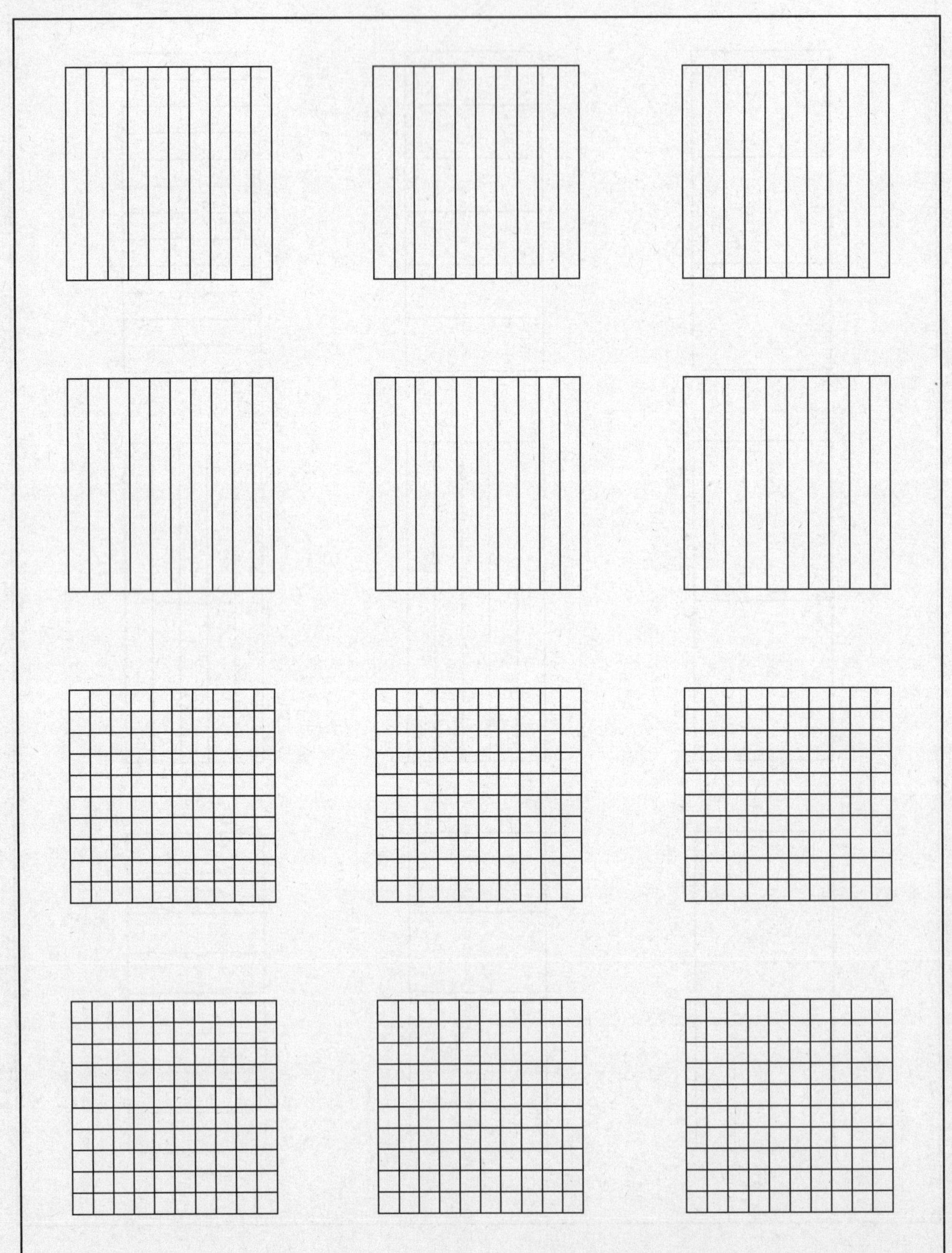

Cut or tear carefully along this line.

Notes

Notes

Notes

Notes

Notes

Notes